The Magic of Numbers

The Magic of Numbers

Benedict Gross

Joe Harris

Harvard University

PEARSON

Prentice Hall

PEARSON EDUCATION, INC.

Upper Saddle River, NJ 07458

Library of Congress Cataloging-in-Publication Data

Gross, Benedict H.
The magic of numbers / Benedict Gross, Joseph Harris.
p. cm.
Includes index.
ISBN 0-13-177721-1
1. Mathematics. I. Harris, Joe. II. Title
QA39.3.G76 2004
510—dc21 2003054832

Executive Acquisition Editor: *Petra Recter*
Editorial/Production Supervision: *Bayani Mendoza de Leon*
Editor in Chief: *Sally Yagan*
Assistant Vice President of Production and Manufacturing: *David W. Riccardi*
Senior Managing Editor: *Linda Mihatov Behrens*
Executive Managing Editor: *Kathleen Schiaparelli*
Assistant Manufacturing Buyer/Manager: *Michael Bell*
Manufacturing Manager: *Trudy Pisciotti*
Marketing Manager: *Krista M. Bettino*
Marketing Assistant: *Annett Uebel*
Director of Creative Services: *Paul Belfanti*
Art Editor: *Jessica Einsig*
Assistant Art Editor: *Abigail Bass*
Creative Director: *Carole Anson*
Art Director: *Jonathan Boylan*
Interior and Cover Designer: *John Christiana*
Cover Manager: *Karen Sanatar*
Editorial Assistant: *Joanne Wendelken*
Art Studio: *Laserwords Private Limited*
Cover Image: *Jasper Johns, "Numbers in Color" 1958–59. Encaustic and collage on canvas. 67 × 49-1/2 in. Albright-Knox Art Gallery, Buffalo, New York. Gift of Seymour H. Knox, Jr., 1959. (c) Jasper Johns/ Licensed by VAGA, New York, NY*

Pearson Prentice Hall
Pearson Education, Inc.
Upper Saddle River, New Jersey 07458

Printed in the United States of America
10 9 8 7 6 5 4 3 2 1

ISBN 0-13-177721-1

Pearson Education LTD., *London*
Pearson Education Australia Pty, Limited, *Sydney*
Pearson Education Singapore, Pte. Ltd.
Pearson Education North Asia Ltd., *Hong Kong*
Pearson Education Canada, Ltd., *Toronto*
Pearson Educación de Mexico, S.A. de C.V.
Pearson Education Japan, *Tokyo*
Pearson Education Malaysia, Pte. Ltd.

“We shall not cease from exploration
And the end of all our exploring
Will be to arrive where we started
And know the place for the first time.”

—T.S. Eliot
“Little Gidding” (Four Quartets)

CONTENTS

Chapter 5

Probability 46

Chapter 6

Pascal's Triangle and the Binomial Theorem 59

Chapter 7

Advanced Counting 69

PART II ARITHMETIC 87

Chapter 8

Divisibility 89

Chapter 9

Chapter 10

Chapter 11

Chapter 12

Chapter 13

Chapter 23

Chapter 24

PREFACE

The primarily purpose of the preface to a textbook is to convey a sense of the goals of the book, and to a lesser extent its level, pace and language. Since this book is based on a course, Quantitative Reasoning 28, that we developed at Harvard University, it seems reasonable to start by describing the goals of that course.

There are, it seems to us, two strains in our educational system, reflecting two disparate aims. One is preprofessional: among the courses you take in high school and in college are those that will, it's hoped, provide you with the basis of the discipline that will become your vocation in later life. The other strain, by contrast, seeks to enrich your life: to expose you to ideas and modes of thought that you might not come in contact with otherwise. This is one of the ideas underlying Harvard's Core Program, and QR28 was developed as a Core course. It's not a technical course, designed to prepare you for the next course; rather, it's simply a collection of topics that we find fascinating, and that make up a coherent whole. Our hope is that we will be able to communicate to you some idea of the mathematical view of the world, and of what attracts people to math in the first place.

Probably the best way to describe the course is by analogy: you might think of it as a math appreciation course, to be taken in the same spirit as you would a music appreciation course. We're not trying to teach you how to write a symphony, or to play the violin; we simply want you to be able to hear the music.

Or you might think of it like an introductory language course—say, Italian—that you take for fun and because it's such a beautiful language. This analogy is particularly apt in one respect. The heart of a language course is not the memorization of a lot of vocabulary and verb tenses—though there's inevitably a lot of that involved—but rather the experience of thinking and speaking in a different tongue. In the same way, in this text there are of necessity a fair number of techniques to learn and calculations to carry out, but that's just the means to an end: our goal, ultimately, is to give you the experience of thinking in math.

What sort of prerequisites does this book have? Well, the technical answer to that is "virtually none": junior high school algebra will cover it handily. (To be concrete, if you can add fractions, and are reasonably comfortable with the use of letters to stand for numbers, you should be solid.) Probably more important, though, is a less quantifiable requirement: we would ask that the reader be prepared to approach the book in a spirit of adventure and exploration, and with the understanding that, while some work will be required, the experience will be worth it.

Additional exercises, problems, and sample exams are available at www.prenhall.com/gross.

Acknowledgments

We are indebted to a great number of people for help in creating this book. Susan Milano gave us the name for the course and, by extension, the book, and also read and critiqued an early version. Sarah Brelsfoard and Debby Green also did a wonderful job of reading the first draft and pointing out mistakes. We owe a debt to Ivan Niven and his book, *Mathematics of Choice*, which has clearly influenced our treatment of the topics in the first part of this book. We would also like to thank the reviewers, Robert Bernhardt of East Carolina University, Fernando Gouvêa of Colby College, Steven Krantz of Washington University, and M. Terrell of Cornell University, for many helpful comments and corrections. And our editors, particularly Erin Mulligan, Bayani Mendoza de Leon, Petra Recter, and Sally Yagan, have been terrifically helpful, as has our compositor at Laserwords.

Inasmuch as this book represents a print version of the course QR28, it could not exist without the help of all those who made the course what it is. Susan Lewis, the director of the Core Program at Harvard, first encouraged us to develop a course in mathematics for this program. Then there are all the people who helped shape the course: our Head Teaching Fellows Tom Weston, Rob Pollack, Elena Mantovan and Nick Rogers; our other Teaching Fellows Sam Williams, Tomas Klenke, Stephanie Yang, Laura DeMarco, Mark Lucianovic, Sarah Dean, Robert Neel and Marty Weissman. And, of course, all the students who took the course from 1999 through 2002.

Benedict Gross

Joe Harris

PART I

Counting

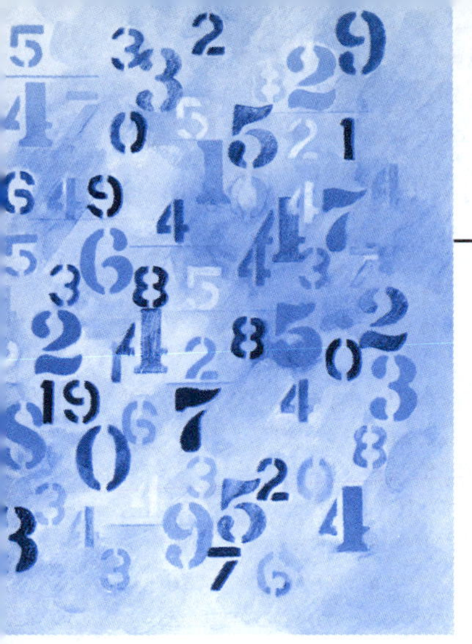

Chapter 1

Simple Counting

It's hard to begin a math book. A few chapters in, it gets easier: by then, writer and reader have—or think they have—a common sense of the level of the book, its pace, language and goals; at that point, communication naturally flows more smoothly. But getting started is awkward.

As a consequence, it's standard practice in math textbooks to include a throw-away chapter or two at the beginning. These often have little or no content; rather, they're put there in the hope of establishing basic terminology and notation, and getting the reader used to the style of the book, before launching into the actual material. Unfortunately, the effect may be the opposite: a chapter full of seemingly obvious statements, expressed in vague language, can have the effect of making the reader generally uneasy without actually conveying any useful information.

Well, far be it from us to deviate from standard practice! The following is our introductory chapter. But here's the deal: you can skip it if you find the material too easy. Really. Just go right ahead to Chapter 2 and start there.

1.1 Counting Numbers

This is a book about numbers. We hope to show you, during its course, something of the wild beauty of numbers: the intricate patterns of their behavior, and the way that even simple operations on them can give rise to questions that people have wrestled with for centuries.

To start things off, we'd like to talk about counting, because that's how numbers first entered our world. It was four or five thousand years ago that people first developed the concept of numbers, probably in order to quantify their possessions and make transactions—my three pigs for your two cows and the like. And the remarkable thing that people discovered about numbers is that the same system of numbers—1, 2, 3, 4, and so on—could be used to count anything: beads, bushels of grain, people living in a village, forces in an opposing army. Numbers can count anything: numbers can even count numbers.

And that's where we'll start. The first problem we're going to pose is simply: how many numbers are there between 1 and 10?

At this point you may be wondering if it's too late to get your money back for this book. Bear with us! We'll get to stuff you don't know soon enough. In the meantime, write them out and count:

$$1, \quad 2, \quad 3, \quad 4, \quad 5, \quad 6, \quad 7, \quad 8, \quad 9, \quad 10$$

—there are 10. How about between 1 and 11? Well, that's one more, so there are 11. Between 1 and 12? 12, of course.

Well, that seems pretty clear; and if we now asked you, for example, how many numbers there are between 1 and 57, you wouldn't actually have to write them out and count; you'd figure (correctly) that the answer would be 57.

OK, then, let's ramp it up a notch. Suppose we ask now: how many numbers are there between 28 and 83, inclusive? ("Inclusive" means that, as before, we include both 28 and 83 in the count.) Well, you could do this by making a list of the numbers between 28 and 83 and counting them, but you have to believe there's a better way than that.

Here's one: suppose you did write out all the numbers between 28 and 83:

$$28, \quad 29, \quad 30, \quad 31, \quad 32, \dots, 82, \quad 83.$$

(Here the dots mean to imagine that we've written all the numbers in between in an unbroken sequence. We'll use this convention when it's not possible or desirable to write out a sequence of numbers in full.) Now subtract the number 27 from each of them. The list now starts at 1, and continues up to $83 - 27 = 56$:

$$1, \quad 2, \quad 3, \quad 4, \quad 5, \dots, 55, \quad 56.$$

From what we just saw we know there are 56 numbers on this list; so there were 56 numbers on our original list as well.

It's pretty clear also that we could do this to count any string of numbers. For example, if we asked how many numbers there are between 327 and 573, you could similarly imagine the numbers all written out:

$$327, \quad 328, \quad 329, \quad 330, \quad 331, \dots, 572, \quad 573.$$

Next, subtract the number 326 from each of them; we get the list

$$1, \quad 2, \quad 3, \quad 4, \quad 5, \dots, 246, \quad 247$$

and so we can conclude that there were $573 - 326 = 247$ numbers on our original list.

Now, there's no need to go through this process every time. It makes more sense to do it once with letters standing for arbitrary numbers, and in that way work out a formula that we can use every time we have such a problem. So: imagine that we're given two whole numbers n and k, with n the larger of the two, and we're asked: how many numbers are there between k and n, inclusive?

We do this just the same way: imagine that we've written out the numbers from k to n in a list

$$k, \quad k+1, \quad k+2, \quad k+3, \dots, n-1, \quad n$$

and subtract the number $k - 1$ from each of them to arrive at the list

$$1, \quad 2, \quad 3, \quad 4, \dots, n-1-(k-1), \quad n-(k-1).$$

Now we know how many numbers are on the list: it's $n - (k - 1)$ or, more simply, $n - k + 1$.[1] Our conclusion, then, is that

> The number of whole numbers between k and n inclusive is
> $$n - k + 1.$$

So, for example, if someone asked "How many numbers are there between 342 and 576?" we wouldn't have to think it through from scratch: the answer is $576 - 342 + 1$, or 235.

Since this is our first formula, it may be time to bring up the whole issue of the role of formulas in math. As we said, the whole point of having a formula like this is that we shouldn't have to recreate the entire argument we used in the concrete examples above every time we want to solve a similar problem. On the other hand, it's also important to keep some understanding of the process, and not to treat the formula as a "black box" that spews out answers. Knowing how the formula was arrived at helps us to know both when it's applicable, and how it can be modified to deal with other situations.

1.2 Counting Divisible Numbers

Now that we've done that, let's try a slightly different problem: suppose we ask "How many even numbers are there between 46 and 104?"

In fact, we can approach this the same way: imagine that we did make a list of all even numbers, starting with 46 and ending with 104:

$$46, \quad 48, \quad 50, \quad 52, \ldots, 102, \quad 104$$

Now, we've just learned how to count numbers in an unbroken sequence. And we can convert this list to just such a sequence if we just divide all the numbers on the list by 2: doing that, we get the sequence

$$23, \quad 24, \quad 25, \quad 26, \ldots, 51, \quad 52$$

of all whole numbers between $46/2$, or 23, and $104/2$, or 52. Now, we know by the formula we just worked out how many numbers there are on that list: there are

$$52 - 23 + 1 = 30$$

numbers between 23 and 52, so we conclude that there are 30 even numbers between 46 and 104.

One more example of this type: let's ask the question, "How many numbers between 50 and 218 are divisible by 3?" Once more we use the same approach: imagine that we made a list of all such numbers. But notice that 50 isn't the first such number, since 3 doesn't divide 50 evenly: in fact, the smallest number on our list that is divisible by 3 is $51 = 3 \times 17$. Likewise, the last number on our list is 218,

---

[1] Is it obvious that $n - (k - 1)$ is the same as $n - k + 1$? If not, take a moment out and convince yourself: subtracting $k - 1$ is the same as subtracting k and then adding 1 back. In this book we'll usually carry out operations like this without comment, but you should take the time to satisfy yourself that they make sense.

which isn't divisible by 3. The largest number on our list which is divisible by 3 is 216, which is ($216 = 3 \times 72$). So the list of numbers divisible by 3 would look like

$$51, \quad 54, \quad 57, \quad 60, \ldots, 213, \quad 216.$$

Now we can do as we did before, and divide each number on this list by 3. We arrive at the list

$$17, \quad 18, \quad 19, \quad 20, \ldots, 71, \quad 72$$

of all whole numbers between 17 and 72, and there are

$$72 - 17 + 1 = 56$$

such numbers.

Now it's time to stop reading for a moment and do some yourself:

Exercise 1.2.1

1. How many numbers between 33 and 97 are even?
2. How many numbers between 17 and 783 are divisible by 6?
3. How many numbers between 45 and 93 are odd?

1.3 "I've reduced it to a previously solved problem"

Note one thing about the sequence of problems we've just done. We started with a pretty mindless one—the number of numbers between 1 and n—which we could answer more or less by direct examination. The next problem we took up—the number of numbers between k and n—we solved by shifting all the numbers down to whole numbers between 1 and $n-k+1$. In effect we reduced it to the first problem, whose answer we knew. Finally, when we asked how many numbers between two numbers were divisible by a third, we answered the question by dividing all the numbers, to reduce the problem to counting numbers between k and n.

This approach—building up our capacity to solve problems by reducing new problems to ones we've already solved—is absolutely characteristic of mathematics. We start out slowly, and gradually accumulate a body of knowledge and techniques; the goal is not necessarily to solve each problem directly, but to reduce it to a previously solved problem.

There's even a standard joke about this:

A mathematician walks into a room. In one corner, he sees an empty bucket. In a second corner, he sees a sink with a water faucet. And, in a third corner, he sees a pile of papers on fire. He leaps into action: he picks up the bucket, fills it up at the faucet, and promptly douses the fire.

The next day, the same mathematician returns to the room. Once more, he sees a fire in the third corner, but this time sitting next to it there's a full bucket of water. Once more he leaps into action: he picks up the bucket, drains it into the sink, places it empty in the first corner and leaves, announcing, "I've reduced it to a previously solved problem!"

Well, maybe you had to be there. But there is a real point to be made here. It's simply this: the ideas and techniques developed in this book are cumulative, each one resting on the foundation of the ones that have come before. We'll occasionally go off on tangents and pursue ideas that won't be used in what follows, and we'll try to tell you when that occurs. But for the most part, *you need to keep up*: that is, you need to work with the ideas and techniques in each section until you feel genuinely comfortable with them, before you go on to the next.

It's worth remarking also that the cumulative nature of mathematics in some ways sets it apart from other fields of science. The theories of physics, chemistry, biology and medicine we subscribe to today flatly contradict those held in the 17th and 18th centuries—it's fair to say that the medical texts dealing with the proper application of leeches are of interest primarily to historians, and we'd bet your high school chemistry course didn't cover phlogiston.[2] By contrast, the mathematics developed at that time is the cornerstone of what we're doing today.

1.4 Really Big Numbers

As long as we're talking about the origins of numbers, let's talk about another important early development: the capacity to write down really big numbers. Think about it: once you've developed the concept of numbers, the next step is to figure out a way to write them down. Of course, you can just make up an arbitrary new symbol for each new number, but this is inherently limited: you can't express large numbers without a cumbersome dictionary.

One of the first treatises ever written on the subject of numbers and counting was by Archimedes, who lived in Syracuse (part of what was then the Greek empire) in the 3rd century B.C. The paper, entitled *The Sand Reckoner*, was addressed to a local monarch, and in it Archimedes claimed that he had developed a system of numbers that would allow him to express as large a number as the number of grains of sand in the universe—a revolutionary idea at the time.

What Archimedes had developed was similar to what we would call *exponential notation*. We'll try to illustrate this by expressing a really large number—say, the approximate number of seconds in the lifetime of the universe. (Don't laugh: this number will actually come up in a practical context in the last part of this book.)

The calculation is simple enough. There are 60 seconds in a minute, and 60 minutes in an hour, so the number of seconds in an hour is

$$60 \times 60 = 3{,}600.$$

There are in turn 24 hours in a day, so the number of seconds in a day is

$$3{,}600 \times 24 = 86{,}400;$$

and since there are 365 days in a (nonleap) year, the number of seconds in a year is

$$86{,}400 \times 365 = 31{,}536{,}000.$$

Now, in exponential notation, we would say this number is roughly 3 times 10 to the 7th power—that is, a 3 with seven 0s after it. (A better approximation, of course,

[2]In case you're curious, phlogiston was the hypothetical principle of fire, of which every combustible substance was in part composed—at least until the whole theory was discredited by Antoine Lavoisier between 1770 and 1790.

would be to say the number is roughly 3.1×10^7, or 3.15×10^7; but we're going to go with the simpler estimate 3×10^7.)

Exponential notation is particularly convenient when it comes to multiplying large numbers. Suppose, for example, that we have to multiply $10^6 \times 10^7$. Well, 10^6 is just $10 \times 10 \times 10 \times 10 \times 10 \times 10$, and 10^7 is just $10 \times 10 \times 10 \times 10 \times 10 \times 10 \times 10$, so when we multiply them we just get the product of 10 with itself 13 times: that is,

$$10^6 \times 10^7 = 10^{13}.$$

In other words, we simply add the exponents. So it's easy to take products of quantities that you've expressed in exponential notation.

For example, to take the next step in our problem, we have to say how old the universe is. Now, that very much depends on your model of the universe. Most astrophysicists believe that the universe is approximately 13.7 billion years old, with a possible error on the order of 1%. We'll write the age of the universe, accordingly, as

$$13{,}700{,}000{,}000 = 1.37 \times 10^{10}$$

years. So the number of seconds in the lifetime of the universe would be approximately

$$(1.37 \times 10^{10}) \times (3 \times 10^7) = 4.11 \times 10^{17};$$

or, rounding it off, the universe is 4×10^{17} seconds old.

You see how we can use this notation to express arbitrarily large numbers. For example, computers currently can carry out on the order of 10^{12} operations a second (a *teraflop*, as it's known in the trade). We could ask: if such a computer were running from the dawn of time to the present, how many operations could it have performed? The answer is, approximately

$$10^{12} \times (4 \times 10^{17}) = 4 \times 10^{29}.$$

Now, for almost all of this book, we'll be dealing with much smaller numbers than these; and we'll be doing exact calculations rather than approximations. But occasionally we will want to express and estimate larger numbers like these. (The last number above—the number of operations a computer running for the lifetime of the universe could perform—will actually arise later on in this book: we'll encounter mathematical processes that require more than this number of operations to carry out.) It's nice to know that we have a notation that can accommodate it.

1.5 It Could Be Worse

Look: this is a math book. We're trying to pretend it isn't, but it is. That means that it'll have jargon—we'll try to keep it to a minimum, but we can't altogether avoid using defined terms. That means that you'll encounter the odd mathematical formula here and there. That means it'll have long discussions aimed at solving artificially posed problems, subject to seemingly arbitrary hypotheses. Mathematics texts have a pretty bad reputation, and we're sorry to say it's largely deserved.

Just remember: it could be worse. You could, for example, be reading a book on Kant. Now, Immanuel Kant is a towering figure in Western Philosophy, a seminal genius who shaped much of modern thought. "The foremost thinker of the Enlightenment and one of the greatest philosophers of all time," the Encyclopedia Britannica calls him. But just read a sentence of his writing:

> *If we wish to discern whether anything is beautiful or not, we do not refer the representation of it to the Object by means of understanding with a view to cognition, but by means of the imagination (acting perhaps in conjunction with understanding) we refer the representation to the Subject and its feeling of pleasure or displeasure.*

What's more, this is not a nugget unearthed from deep within one of Kant's books. It is, in fact, the first sentence of the first Part of the first Moment of the first Book of the first Section of Part I of Kant's *The Critique of Judgement*.

Now, we're not trying to be anti-intellectual here, or to take cheap shots at other disciplines. Just the opposite, in fact: what we're trying to say is that any body of thought, once it progresses past the level of bumper sticker catch phrases, requires a language and a set of conventions of its own. These provide the precision and universality that are essential if people are to communicate and develop the ideas further, and shape them into a coherent whole. But they also can have the unfortunate effect of making much of the material inaccessible to a casual reader. Mathematics suffers from this—as do most serious academic disciplines.

The point, in other words, is not that the passage from Kant we just quoted is babble; it's not. (Lord knows we could have dug up enough specimens of academic writing that are, if that was our intention.) In fact, it's the beginning of a serious and extremely influential attempt to establish a philosophical theory of aesthetics. As such, it may be difficult to understand without some mental effort. It's important to bear in mind that the apparent obscurity of the language is a reflection of this difficulty, not necessarily the cause of it.

So, the next time you're reading this book and you encounter a term that turns out to have been defined—contrary to its apparent meaning—some 30 pages earlier, or a formula that seems to come out of nowhere and that you're apparently expected to find self-explanatory, just remember: it could be worse.

© 1999 Bill Amend/Dist. by Universal Press Syndicate
WE HAD A MARINE CORPS RECRUITER TALK TO OUR CLASS TODAY.
www.foxtrot.com
HE TOLD US ALL ABOUT THE RIGORS OF BOOT CAMP: THE 4 A.M. WAKE-UP CALLS...THE TWENTY-MILE RUNS IN FULL COMBAT GEAR...THE OBSTACLE COURSES WITH BARBED WIRE AND LIVE AMMO...
12-6 AMEND
THAT DOESN'T SOUND TOO APPEALING.
THEN HE HELD UP FOR COMPARISON A COLLEGE CALCULUS TEXTBOOK.
SMART MAN.
HE COULDN'T HAND OUT APPLICATIONS FAST ENOUGH.

Chapter 2

The Multiplication Principle

2.1 Choices

Let's suppose you climb out of bed one morning, still somewhat groggy from the night before. You grope your way to your closet, where you discover that your cache of clean clothes has been reduced to four shirts and three pairs of pants. It's far too early to exercise any aesthetic judgment whatsoever: any shirt will go with any pants; you only need something that will get you as far as the dining hall and that blessed, life-giving cup of coffee. The question is,

How many different outfits can you make out of your four shirts and three pairs of pants?

Admittedly the narrative took a sharp turn toward the bizarre with that last sentence. Why on earth would you or anyone care how many outfits you can make? Well, bear with us while we try to answer it anyway.

Actually, if you thought about the question at all, you probably have already figured out the answer: each of the four shirts is part of exactly three outfits, depending on which pants you choose to go with it, so the total number of possible outfits is $3 \times 4 = 12$. (Or, if you like to get dressed from the bottom up, each of the three pairs of pants is part of exactly four outfits; either way the answer is 3×4.) If we're feeling really fussy, we could make a table: say the four shirts are a golf shirt, an oxford, a tank top and a T-shirt extolling the virtues of your favorite athletic wear, and the pants consist of a pair of jeans, some cargo pants and a pair of shorts. Then we can arrange the outfits in a rectangle:

golf shirt w/ jeans	oxford shirt w/ jeans	tank top w/ jeans	T-shirt w/ jeans
golf shirt w/ cargo pants	oxford shirt w/ cargo pants	tank top w/ cargo pants	T-shirt w/ cargo pants
golf shirt w/ shorts	oxford shirt w/ shorts	tank top w/ shorts	T-shirt w/ shorts

Now, you know we're not going to stop here. Suppose next that, in addition to picking a shirt and a pair of pants, you also have to choose between two pairs of shoes. Now how many outfits are there?

Well, the idea is pretty much the same: for each of the possible shirt/pants combinations, there are two choices for the shoes, so the total number of outfits is $4 \times 3 \times 2 = 12 \times 2 = 24$. And if in addition we had a choice of five hats, the total number of possible outfits would be $4 \times 3 \times 2 \times 5 = 120$—you get the idea.

Now it's midday and you head over to the House of Pizza to order a pizza for lunch. You feel like having one meat topping and one vegetable topping on your pizza; the House of Pizza offers you seven meat toppings and four vegetable toppings. How many different pizzas do you have to choose among?

"That's the same problem with different numbers!" you might say, and you'd be right: to each of the seven meat toppings you could add any one of the four vegetable toppings, so the total number of different pizzas you could order would be 7×4, or 28.

Evening draws on, and your roommates send you out to the local House of Videos to rent some videos. You're going to have a triple feature in your room: one action film, one lighthearted romantic comedy and one movie based on a cartoon or video game. The House of Videos, following its corporate plan, has in stock a thousand copies each of seven action movies, five lighthearted romantic comedies and 23 movies based on cartoons or video games. How many triple features can you rent?

"That's the same problem again!" you might be thinking: the answer's just the number 7 of action movies times the number 5 of lighthearted romantic comedies times the number 23 of movies based on a cartoon or video game, or $7 \times 5 \times 23 = 805$. Trust us!—we are going somewhere with this. But you're right, it's time to state the general rule that we're working toward, which is called the *multiplication principle*:

> The number of ways of making a sequence of independent choices is the product of the number of choices at each step.

Here "independent" means that how you make the first choice doesn't affect the number of choices you have for the second, and so on. In the first case above, for example—getting dressed in the morning—it corresponds to having no fashion sense whatsoever.

The multiplication principle is easy to understand and apply, but awkward to state in reasonably coherent English, which is why we went through three examples before announcing it. In fact, you may find the examples more instructive than the principle itself; if the boxed statement seems obscure to you, just remember: "4 shirts, 3 pants, 12 outfits."

2.2 More Examples

An old-style Massachusetts license plate has on it a sequence of three numbers followed by three letters. How many different old-style Massachusetts license plates can there be?

This is easy enough to answer: we have 10 choices for each of the numbers, and 26 choices for each of the letters; and since none of these choices is constrained in any way, the total number of possible license plates is

$$10 \times 10 \times 10 \times 26 \times 26 \times 26 = 17{,}576{,}000$$

A similar question is this. Suppose for the moment that by "word" we mean any sequence of the 26 letters of the English alphabet—we're not going to make a distinction between actual words and arbitrary sequences. How many three-letter words are there?

This is just the same as the license plate problem (or at least the first half): we have 26 independent choices for each of the letters, so the number of three-letter words is $26^3 = 17{,}576$. In general,

$$\# \text{ of 1-letter words} = 26$$

$$\# \text{ of 2-letter words} = 26^2 = 676$$

$$\# \text{ of 3-letter words} = 26^3 = 17{,}576$$

$$\# \text{ of 4-letter words} = 26^4 = 456{,}976$$

$$\# \text{ of 5-letter words} = 26^5 = 11{,}881{,}376$$

$$\# \text{ of 6-letter words} = 26^6 = 308{,}915{,}776$$

and so on.

Next, let's suppose that there are 15 students in a class, and that they've decided to choose a set of class officers: a president, a vice president, a secretary and a treasurer. How many possible slates are there? That is, how many ways are there of choosing the four officers?

Actually, there are two versions of this question, depending on whether or not a single student is allowed to hold more than one of the positions. If we assume first that there's no restriction of how many positions one person can hold, the problem is identical to the ones we've just been looking at: we have 15 choices each for the four offices, and they are all independent, so that the total number of possible choices is

$$15 \times 15 \times 15 \times 15 = 50{,}625.$$

Now suppose on the other hand we impose the rule that no person can hold more than one office. How many ways are there of choosing officers?

Well, this can also be done by the multiplication principle. We start (say) by choosing the president; we have clearly 15 choices there. Next, we choose the vice president. Now our choice is restricted by the fact that our newly selected president is no longer eligible, so that we have to choose among the 14 remaining students. After that we choose a secretary, who could be anyone in the class except the two officers already chosen; so we have 13 choices here; and finally we choose a treasurer from among the 12 students in the class other than the president, vice president and secretary. Altogether, the number of choices is

$$15 \times 14 \times 13 \times 12 = 32{,}760.$$

Note one point here: in this example, the actual choice of, say, the vice president *does* depend on who we chose for president; the choice of a secretary does depend on who we selected for president and vice president, and so on. But the *number* of choices doesn't depend on our prior selections, so the multiplication principle still applies.

In a similar vein, we could modify the question we asked a moment ago about the number of three-letter words, and ask: how many 3 letter words have no repeated letters? The solution is completely analogous to the class-officer problem: we have 26 choices for the first letter, 25 for the second and 24 for the third, so that we have a total of

$$26 \times 25 \times 24 = 15{,}600$$

such words. In general, we can calculate

$$\begin{aligned}
\text{\# of 1-letter words} &= 26 \\
\text{\# of 2-letter words w/o repeated letters} &= 26 \cdot 25 = 650 \\
\text{\# of 3-letter words w/o repeated letters} &= 26 \cdot 25 \cdot 24 = 15{,}600 \\
\text{\# of 4-letter words w/o repeated letters} &= 26 \cdot 25 \cdot 24 \cdot 23 = 358{,}800 \\
\text{\# of 5-letter words w/o repeated letters} &= 26 \cdot 25 \cdot 24 \cdot 23 \cdot 22 = 7{,}893{,}600 \\
\text{\# of 6-letter words w/o repeated letters} &= 26 \cdot 25 \cdot 24 \cdot 23 \cdot 22 \cdot 21 = 165{,}765{,}600
\end{aligned}$$

and so on.

Now, here's an interesting (if somewhat tangential) question. Let's compare the numbers of words of each length to the number of words with no repeated letters. What percentage of all words have repetitions, and what percentage don't? Of course, as the length of the word increases, we'd expect a higher proportion of all words to have repeated letters—relatively few words of two or three letters have repetitions, while of course every word of 27 or more letters does. We could ask, then: when does the fraction of words without repeated letters dip below one-half? In other words, for what lengths do the words with repeated letters outnumber those without?

Before we tabulate the data and give the answer, you might want to take a few minutes and think about the question. What would your guess be?

STOP.

CLOSE THE BOOK.

GRAB A PAD OF PAPER AND A PEN.

WORK OUT SOME EXAMPLES ON YOUR OWN.

THINK.

length	number of words	without repeats	% without repeats
1	26	26	100.00
2	676	650	96.15
3	17,576	15,600	88.76
4	456,976	358,800	78.52
5	11,881,376	7,893,600	66.44
6	308,915,776	165,765,600	53.66
7	8,031,810,176	3,315,312,000	41.28
8	208,827,064,576	62,990,928,000	30.16
9	5,429,503,678,976	1,133,836,704,000	20.88

Now that's bound to be surprising: among six-letter words, those with repeated letters represent nearly half, and among seven-letter words they already substantially outnumber the words without repeats. In general, the percentage of words without repeated letters drops off pretty fast: by the time we get to twelve-letter words, fewer than 1 in 20 has no repeated letter. We'll see another example of this phenomenon when we talk about the birthday problem in Section 5.6.

2.3 Two Formulas

There are two special cases of the multiplication principle that occur so commonly in counting problems that they're worth mentioning on their own, and we'll do that here. Neither will be new to us; we've already encountered examples of each.

Both involve repeated selections from a single pool of objects. If there are no restrictions at all on the choices, the application of the multiplication principle is particularly simple: each choice in the sequence is a choice among all the objects in the collection. If we're counting three-letter words in an alphabet of 26 characters, for example—where by "word" we again mean an arbitrary sequence of letters—there are 26^3; if we're counting four-letter words in an alphabet of 22 characters, there are 22^4; and so on. In general, we have the following rule:

> The number of sequences of k objects chosen from a collection of n objects is n^k.

The second special case involves the same problem, but with a commonly applied restriction: we're again looking at sequences of objects chosen from a common pool of objects, but this time we're not allowed to choose the same object twice. Thus, the first choice is among all the objects in the pool; the second choice is among all but one, the third among all but two, and so on; if we're looking at a sequence of k choices, the last choice will be among all but the $k-1$ already chosen. Thus, as we saw, the number of three-letter words without repeated letters in an alphabet of 26 characters is $26 \cdot 25 \cdot 24$; the number of four-letter words without repeated letters in an alphabet of 22 characters is $22 \cdot 21 \cdot 20 \cdot 19$; and so on. In general, if the number of objects in our pool is n, the first choice will be among all n; the second among $n-1$, and so on. If we're making a total of k choices, the last choice will exclude the $k-1$ already chosen; that is, it'll be a choice among the $n-(k-1)=n-k+1$

objects remaining. The total number of such sequences is thus the product of the numbers from n down to $n - k + 1$. We write this as

$$n \cdot (n-1) \cdot (n-2) \cdot \cdots \cdot (n-k+1),$$

where the dots in the middle indicate that you're supposed to keep going multiplying all the whole numbers in the series starting with n, $n - 1$ and $n - 2$, until you get down to $n - k + 1$. Time for a box:

> The number of sequences of k objects chosen without repetition from a collection of n objects is
>
> $$n \cdot (n-1) \cdot (n-2) \cdot \cdots \cdot (n-k+1).$$

Exercise 2.3.1 In one of the Massachusetts state lotteries, the winning number is chosen by picking six ping-pong balls from a bin containing balls labeled "1" through "36" to arrive at a sequence of six numbers between 1 and 36. Ping-pong balls are not replaced after they're chosen; that is, no number can appear twice in the sequence. How many possible outcomes are there?

Note that in this last exercise, the order in which the ping-pong balls are chosen is relevant: if the winning sequence is "17-32-5-19-12-27" and you picked "32-17-5-19-12-27," you *don't* get to go to work the next day and tell your boss what you really think of her.

Exercise 2.3.2 The Hebrew alphabet has 22 letters. How many five-letter words are possible in Hebrew? (Again, by "word" we mean just an arbitrary sequence of five characters from the Hebrew alphabet.) What fraction of these have no repeated letters?

2.4 Factorials

The two formulas we described in the last section were both special cases of the multiplication principle. There is in turn a special case of the second formula that crops up fairly often and that's worth talking about now. We'll start, as usual, with an example.

EXAMPLE 2.4.1 Suppose that we have a first-grade class of 15 students, and we want to line them up to go out to recess. How many ways of lining them up are there—that is, in how many different orders can they be lined up?

SOLUTION Well, think of it this way: we have 15 choices of who'll be first in line. Once we've chosen the line leader, we have 14 choices for who's going to be second, 13 choices for the third, and so on. In fact, all we're doing here is choosing a sequence of 15 children from among the 15 children in the class, without repetition; whether we invoke the formula in the last section or do it directly, the answer is

$$15 \cdot 14 \cdot 13 \cdot 12 \cdot 11 \cdot 10 \cdot 9 \cdot 8 \cdot 7 \cdot 6 \cdot 5 \cdot 4 \cdot 3 \cdot 2 \cdot 1 = 1{,}307{,}674{,}368{,}000,$$

or about 1.3×10^{12}—more than a trillion orderings. ■

In general, if we ask how many ways there are of placing n objects in a sequence, the answer is the product of all the whole numbers between 1 and n. This is a quantity that occurs so often in mathematics (and especially in counting problems) that it has its own symbol and name:

> The product $n \cdot (n-1) \cdot (n-2) \cdot \cdots \cdot 3 \cdot 2 \cdot 1$ of the numbers from 1 to n is written $n!$ and called "n factorial."

Here's a table of the factorials up to 15:

n	$n!$
1	1
2	2
3	6
4	24
5	120
6	720
7	5,040
8	40,320
9	362,880
10	3,628,800
11	39,916,800
12	479,001,600
13	6,227,020,800
14	87,178,291,200
15	1,307,674,368,000

There are many fascinating things to be said about these numbers. Their size alone is an interesting question: we've seen that 15 factorial is over a trillion; approximately how large a number is, say, 100 factorial? But we'll leave these questions aside for now. At this point, we'll be using factorials for the most part just as a way of simplifying notation. We'll start with the last formula of the preceding section.

It's pretty obvious that writing 15! is a whole lot easier than writing out the product $15 \cdot 14 \cdot 13 \cdot 12 \cdot 11 \cdot 10 \cdot 9 \cdot 8 \cdot 7 \cdot 6 \cdot 5 \cdot 4 \cdot 3 \cdot 2 \cdot 1$. But there are other, less obvious uses of the notation. Suppose, for example, that we wanted to make up a baseball team out of the 15 kids in the class—that is, choose a sequence of nine of the kids in the class of 15, without repetition. We'd have 15 choices for the pitcher, 14 for the catcher, 13 for the first baseman, and so on. When you choose the ninth and last player, you'll be choosing among the $15 - 8 = 7$ kids left at that point, so that the total number of teams would be

$$15 \cdot 14 \cdot 13 \cdot 12 \cdot 11 \cdot 10 \cdot 9 \cdot 8 \cdot 7.$$

But there's a faster way to write this number, using factorials. Basically, we could think of this product as the product of all the numbers from 15 down to 1, except we leave off the numbers from 6 down to 1—in other words, the product of the numbers from 15 to 1 divided by the product of the numbers from 6 to 1, or

$$\frac{15!}{6!}.$$

Now, this may seem like a strange way of writing out the product: it seems inefficient to multiply all the numbers from 15 to 1 and then divide by the product of the numbers you didn't want in the first place. And it is—no one in his right mind would calculate out the number that way. But just as notation, "$\frac{15!}{6!}$" takes up a whole lot less space than "$15 \cdot 14 \cdot 13 \cdot 12 \cdot 11 \cdot 10 \cdot 9 \cdot 8 \cdot 7$," and we'll go with it. For example, we'll rewrite the boxed formula from the last section:

> The number of sequences of k objects chosen without repetition from a collection of n objects is $n!/(n-k)!$

One final note about factorial notation: it is the standard convention that $0! = 1$. You could think of this as the answer to a Zen koan: "How many ways are there of ordering no objects?" But we'll ignore the philosophical ramifications here and simply accept it as a notational convention: it just makes the formulas come out better, as we'll see.

2.5 Another Wrinkle

The multiplication principle itself is completely straightforward. But sometimes there may be more than one way to apply it; and sometimes one of those ways will work when another doesn't. We have to be prepared, in other words, to be flexible in applying the multiplication principle. We'll see lots of examples of this over the course of this part of the book; here's one of them.

To start with, let's take a simple problem: how many three-digit numbers can you form using the numbers 1 through 9, with no repeated digit? As we've seen already, this is completely straightforward: we have nine choices for the first digit, then eight choices for the second and finally seven choices for the third, for a total of

$$9 \times 8 \times 7 = 504$$

choices.

Now let's change the problem a bit: suppose we ask, "How many of those 504 numbers are odd?" In other words, how many have as their third digit a 1, 3, 5, 7 or 9?

We can try to do it the same way: as before, there are nine choices for the first digit and eight for the second. But when we get to the third digit, we're stuck. For example, if the first two digits we selected were 2 and 4, then the third digit could be any of the numbers 1, 3, 5, 7 or 9, so we have five choices. If the first two digits were 5 and 7, however, the third digit could only be a 1, 3 or 9; we have only three choices. The choices, in other words, don't seem to be independent.

But they are if we make them in a different order! Suppose that rather than choosing the first digit first and so on, we go from right to left instead—in other words, choose the third digit first, then the middle and finally the first. Now we can choose the last digit freely among the numbers 1, 3, 5, 7 and 9, for a total of five choices. The choice of the middle digit is constrained only by the requirement that it not repeat the one we've already chosen; so there are eight choices for it, and likewise seven choices for the first digit. There are thus

$$5 \times 8 \times 7 = 280$$

such numbers.

Sometimes we find ourselves in situations where the multiplication principle may not seem applicable, but in fact its application is completely straightforward as long as we keep our wits about us. Here's an example:

EXAMPLE 2.5.1 Suppose that in the class we were discussing in Example 2.4.1 there are eight boys and seven girls, and we want to line them up so that no two boys are next to each other. How many ways are there of doing this?

SOLUTION Actually, before we go and give the solution, let's take a moment and see that the multiplication principle fails. In fact, if we try to use the same approach as we took to in solving Example 2.4.1, it screws up already at the second step. That is, we seemingly have as before 15 choices of who's to be first in line. But the numbers of possible choices for who goes second depends on our first choice: if we chose a girl to be first, there are no restrictions on who goes second, and there are 14 choices; but if we chose a boy to be first in line, the second in line must be chosen from among the seven girls.

We need, in other words, a different approach. But here we're in luck: if we think about it, we can see that since there are 8 boys out of 15 kids, and no two boys are to be next to each other in line, the line must alternate boy/girl/boy/girl until the final place, which must be a boy. In other words, the odd-numbered places in line must all be occupied by boys, and the even places by girls.

Thus, to choose an ordering of the whole class subject to the constraint that no two boys are next to each other, we have to choose a first boy, a second boy, and so on until we get to the eighth and last boy; and likewise we have to choose a first girl, a second girl, and so on to the seventh girl. Put another way, we simply have to order the boys and the girls separately. We know that there are 8! ways of ordering the boys and 7! ways of ordering the girls, so the multiplication principle tells us that the total number of ways of lining up the class is

$$8! \cdot 7! = 203{,}212{,}800. \qquad \blacksquare$$

EXAMPLE 2.5.2 One last puzzler: suppose that there were six boys and nine girls, and again we wanted to line up the class so that no two boys are next to each other. How many ways would there be of doing this?

In fact, this is a *much* harder problem, because we can't avail ourselves of the trick we used in the last example. But it is one you'll learn how to do. So think for a while about how you might try to approach it, and in Section 4.3 we promise we'll work it out.

Chapter 3

The Subtraction Principle

There are 26 letters in the English alphabet, divided into vowels and consonants. For the purposes of this discussion, we'll say the letters

A, E, I, O and U

are vowels, and the letters

B, C, D, F, G, H, J, K, L, M, N, P, Q, R, S, T, V, W, X, Y and Z.

are consonants. Now, quickly: *how many consonants are there?*

How many of you counted out the letters in the sequence B, C, D, F, . . .? Probably not many: it's just a lot easier to count the vowels, and subtract the number 5 of vowels from total number 26 of letters to arrive at the answer that there are $26 - 5 = 21$ consonants.

And that's all there is to the subtraction principle, which is the second of the basic counting tools we'll be using, after the multiplication principle. It's not at all deep—it amounts to nothing more than an observation, really—but we'll give it a box anyway:

> The number of objects in a collection that satisfy some condition is equal to the total number of objects in the collection minus the number of those that don't.

The point being, it's often easier to count the latter than the former. It hardly warrants a box of its own, but—in conjunction with the multiplication principle—it gives us a number of different ways of approaching a lot of counting problems. In fact, as we'll see in this chapter and the next, it greatly broadens the scope of the problems we can solve.

We start with a few simple examples . . .

3.1 Back to the Video Store

Once more, you're going to have a triple feature in your room: one action film, one lighthearted romantic comedy and one movie based on a cartoon or video game. The House of Videos, again, has in stock a thousand copies each of seven action movies (five of which feature car chases), five lighthearted romantic comedies (two of which feature car chases) and 23 movies based on cartoons or video games (eight of which feature car chases). But there's one restriction: some of your roommates have informed you that if you return with three movies featuring car chases they're officially kicking you out of the room. Now how many triple bills are possible?

Well, we could try to do this with the multiplication principle, as we did before the anti-car chase faction in your room raised its voice. But it's easy to see this isn't going to work. We can pick the action movie freely, of course; we have seven choices there. And we can pick the lighthearted romantic comedy freely as well; that's a free choice among five movies. But when it comes time to pick the last movie, how many choices we have depends on what our choices up to that point have been: if either of the first two movies is without car chases, we can choose the third movie freely among the 23 movies based on cartoons or video games; but if both of our first two choices do feature car chases, the choice of the third movie is limited to those 15 that don't. Changing the order of selection doesn't help, either: any way we work it, the number of choices available to us for the last movie depends on our first two selections.

So what do we do? It's simple enough. We already know how many total choices we'd have if there were no restrictions: as we worked it out, it's just

$$7 \times 5 \times 23 = 805.$$

At the same time, it's easy enough to count the number of triple bills that are excluded if we want to stay in the room: we can choose any of the five action movies featuring car chases, either of the two lighthearted romantic comedies featuring car chases, and any of the eight movies based on cartoons or video games featuring car chases, for a total of

$$5 \times 2 \times 8 = 80$$

disallowed triple features. The number of allowable choices is thus

$$805 - 80 = 725.$$

Here's a similar problem (some might say the same problem). We've already counted the number of four-letter words, by which we mean arbitrary sequences of four of the 26 characters in the English alphabet. Suppose we ask now, *how many such words have at least one vowel*? (Here we'll stick to the convention that "Y" isn't a vowel.)

As in the last problem, the multiplication principle seems to work fine until we get to the last letter, and then it breaks down. We have 26 choices for the first letter, 26 for the second and 26 for the third. But when it comes to choosing the last letter, we don't know how many choices we'll have: if any of the preceding three choices happened to be a vowel, we are now free to choose any letter for the last one in our word; but if none of the first three was a vowel we can only choose among the five vowels for the last.

Instead, we use the subtraction principle: we know how many words there are altogether, and *we'll subtract from that the number of words consisting entirely of consonants*. Both are easy: the number of all possible words is just 26^4, and the number of four-letter words consisting only of consonants is 21^4, so the answer to our problem is

$$26^4 - 21^4 = 456{,}976 - 194{,}481 = 262{,}495.$$

One more example: in the first section, we saw how to answer questions like, "How many numbers are there between 34 and 78?" and "How many numbers between 34 and 78 are divisible by 5?" Well, suppose now someone asks, "How many numbers between 34 and 78 are *not* divisible by 5?"

It's pretty clear this is a case for the subtraction principle. We know the number of numbers between 34 and 78 is

$$78 - 34 + 1 = 45.$$

Moreover, since the first and last numbers between 34 and 78 that are divisible by 5 are $35 = 7 \times 5$ and $75 = 15 \times 5$, the number of numbers in this range divisible by 5 is the number of numbers between 7 and 15; that is,

$$15 - 7 + 1 = 9.$$

So by the subtraction principle, the number of numbers between 34 and 78 that are not divisible by 5 is $45 - 9$, or 36.

3.2 Some More Problems

With the subtraction principle, we've doubled the number of techniques we can apply to counting problems. One downside to having more than one technique, though, is that it's no longer unambiguous how to go about solving a problem: we may need to use one technique, or the other, or a combination. This is the beginning of the art of counting, and to develop our technique we'll work out a few more examples.

For a start, let's go back to that first-grade classroom we were at before in Example 2.4.1, the one with 15 kids that need to be lined up. This time, though, let's make the problem a little more difficult: let's suppose that two of the kids in the class, Bobby and Jason, are truly obnoxious little brats. Either one individually is unruly to the point of psychopathy; the last thing in the world you'd want is the two of them standing next to each other in line. So, the problem we're going to deal with is,

EXAMPLE 3.2.1 How many ways are there of lining up the class so Bobby and Jason are *not* next to each other?

SOLUTION Well, we can certainly apply the subtraction principle here: we know there are 15! ways there are of ordering the class if we pose no restrictions; so if we can figure out how many ways there are of lining them up so that Bobby and Jason *are* next to each other, we can subtract that from the total and get the answer that way.

So, how do we figure out the number of lineups with Bobby and Jason adjacent? It seems we haven't exactly solved the problem yet: the next thing we see is that the multiplication principle isn't going to work here, at least not as we applied it in Example 2.4.1. We can certainly choose any of the 15 students to occupy the first place in line, but then the number of choices for the second place in line depends on whether the first choice was Bobby or Jason, or one of the other 13 kids. What's more, this ambiguity persists at every stage thereafter: whom we can put in each place in line depends on who we put in the preceding spot.

But there are other ways of applying the multiplication principle in this setting. In the solution we gave to Example 2.4.1, we made our choices one place at a time—that is, we chose one of the 15 kids to occupy the first place in line; then we chose one of the remaining 14 kids to occupy the second place, and so on. But we could have done it the other way around: we could have taken the kids one at a time, and assigned each a place in line. For example, we could start with Bobby, and assign him any of the 15 places in line; then go on to Jason and assign him any of the remaining 14 places in line, and so on through all 15 kids.

As long as we're dealing with the version of the problem given in Example 2.4.1, it doesn't matter which approach we take; both lead us to the answer $15 \cdot 14 \cdot 13 \ldots 3 \cdot 2 \cdot 1 = 15!$. But in the current situation—where we're trying to count the number of lineups with Bobby and Jason adjacent, say—it does make a difference.

At first it may not seem like it. Doing it this way, we can assign Bobby to any of the 15 places in line; but then the number of choices we have for Jason depends on where we assigned Bobby: if Bobby was placed in either the first or the fifteenth place in line, we will have no choice but to place Jason in the second or fourteenth, respectively; but if Bobby was placed in any of the interior slots, then we can choose to place Jason either immediately ahead of him or immediately behind him. So it seems that the multiplication principle doesn't work this way, either.

But there is a difference. Approaching the problem this way—taking the students one at a time, and assigning each in turn one of the remaining places in line—we see that once we've got Bobby and Jason assigned their places, the multiplication principle takes over: there are 13 choices for where to place the next kid, 12 choices of where to place the one after that, and so on. In other words, if we break the problem up into first assigning Bobby and Jason their places, and then assigning the remaining 13 kids theirs, we see that

$$\left\{\begin{array}{l}\text{the number of lineups}\\ \text{of the class with Bobby}\\ \text{and Jason adjacent}\end{array}\right\} = \left\{\begin{array}{l}\text{the number of ways of}\\ \text{assigning Bobby and Jason}\\ \text{adjacent places in line}\end{array}\right\} \times 13!$$

It remains to count the number of ways of assigning Bobby and Jason adjacent places in line. This is not hard: as we saw above, there are two ways of doing this with Bobby occupying an end position, and $13 \times 2 = 26$ ways of doing it with Bobby occupying an interior position (second through fourteenth), for a total of 28 ways. Or we could count this way: to specify adjacent places in line for Bobby and Jason, we could first specify the *pair* of positions they're to occupy—first and second, or second and third, and so on up to fourteenth and fifteenth—and then say which of the pair Bobby's to occupy. For the first, there are 14 choices, and for the latter 2 choices, so by the multiplication principle we see again there are 28 ways of assigning Bobby and Jason adjacent places in line.

In conclusion, we see that

$$\left\{\begin{array}{l}\text{the number of lineups}\\ \text{of the class with Bobby}\\ \text{and Jason adjacent}\end{array}\right\} = 28 \times 13!$$

and correspondingly,

$$\left\{\begin{array}{l}\text{the number of lineups} \\ \text{of the class with Bobby} \\ \text{and Jason apart}\end{array}\right\} = 15! - (28 \times 13!)$$

$$= 1{,}133{,}317{,}785{,}600. \qquad \blacksquare$$

Exercise 3.2.2 Do Example 3.2.1 over, using the approach followed above but without the subtraction principle: that is, count the number of lineups of the class with Bobby and Jason apart by counting the number of ways you can assign Bobby and Jason to two *nonadjacent* places in line, and the number of ways you can assign the remaining 13 students to the remaining 13 places. Does your answer agree with the one above?

Before we go on, we want to emphasize one point that is illustrated by Example 3.2.1 and its solution. It's an important aspect of learning and doing mathematics, and the failure to appreciate it is the cause of a lot of the frustration that everyone experiences in reading math books. Simply put, it's this: *formulas don't work*. At least, they don't usually work in the sense that you can just plug in appropriate numbers, turn the crank and arrive at an answer. It's better to think of formulas as guides, suggesting effective ways of thinking about problems.

That's probably not what you wanted to hear. When it's late at night and your math homework is the only thing standing between you and bed, you don't want to embark on a glorious journey of exploration and discovery. You just want someone to tell you what to do to get the answer, and formulas may appear to do exactly that. But, really, that's not what they're there for, and appreciating that fact will spare you a lot of aggravation.

Now you try it.

Exercise 3.2.3 A new-style Massachusetts license plate has two letters (which can be any letter from A to Z) followed by four numbers (which can be any digits from 0 to 9).

1. How many new-style Massachusetts license plates are there?
2. How many new-style Massachusetts license plates are there if we require no repeated letters and no repeated numbers?
3. How many new-style Massachusetts license plates are there that have at least one 7?

Exercise 3.2.4 Let's assume that a phone number has seven digits, and cannot start with a 0.

1. How many possible phone numbers are there?
2. How many phone numbers are there with at least one even digit?

Exercise 3.2.5 Getting dressed: suppose you own eight shirts, five pairs of pants and three pairs of shoes.

1. Assuming you have no fashion sense whatsoever, how many outfits can you make?

2. Suppose now that one shirt is purple, one pair of pants is red, and you can make any combination *except* ones including the red pants and the purple shirt. How many outfits can you make?
3. Now suppose that any time you wear the purple shirt you *must* also wear the red pants. How many outfits can you make?

The following problem is hard, but doesn't use any ideas that we haven't introduced.

Exercise 3.2.6 Let's go back to the problem of lining up our class of 15 students. Suppose that Bobby and Jason are so wired that for the sake of everyone's sanity we feel there should be at least two other kids between them. Now how many possible lineups are there?

3.3 Multiple Subtractions

Even as simple an idea as the subtraction principle sometimes has complications. In this section, we'll discuss some of what can happen when we have to exclude more than one class of object from a pool. As with the subtraction principle itself, the basic concept is more common sense than arithmetic, and to emphasize that point we'll start with an edible example.

Consider the following list of 17 vegetables:

artichokes
asparagus
beets
broccoli
cabbages
carrots
cauliflower
celery
corn
eggplant
lettuce
onions
peas
peppers
potatoes
spinach
zucchini

Of these, four—beets, carrots, onions and potatoes—are root vegetables. Two—corn and potatoes—are starchy. Now we ask the question: how many are neither root vegetables nor starchy?

Well, the obvious thing to do would be to subtract the number of root vegetables and starchy vegetables from the total, getting the answer

$$17 - 4 - 2 = 11.$$

But a moment's thought (or, for that matter, actual counting) shows you that isn't right: because a potato is both a root vegetable and a starchy one, you've subtracted it twice. The correct answer is accordingly 12.

And that's the point of this section. It amounts to the observation that when you want to exclude two classes of objects from a pool and count the number left, you can start with the total number of objects in the pool and subtract the number of objects in each of the two excluded categories; *but then you have to add back in the number of objects that belong to both classes and have therefore been subtracted twice.*

Here's a more mathematical example:

EXAMPLE 3.3.1 How many numbers between 100 and 1,000 are divisible by neither 2 nor 3?

SOLUTION We know that the number of numbers between 100 and 1,000 is simply

$$1{,}000 - 100 + 1 = 901.$$

Likewise, we can count the numbers in this range divisible by 2: these are just the even numbers between 100 and 1,000, or in other words twice the numbers between 50 and 500; so there are

$$500 - 50 + 1 = 451$$

of them. Similarly, the numbers divisible by 3 are just 3 times the numbers between 34 and 333; so there are

$$333 - 34 + 1 = 300$$

of those. So, naively, we want to subtract each of 451 and 300 from the total 901.

But, as you've probably figured out—we've stepped all over this punchline—that would be wrong. Because there are numbers divisible by both 2 and 3, and these will have been subtracted twice; to rectify the count we have to add them back in once.

Now, what numbers are divisible by both 2 and 3? The answer is that a number divisible by 2 and by 3 is necessarily divisible by 6, and vice versa.[1] So the numbers between 100 and 1,000 that are divisible by both 2 and 3 are just the numbers in that range divisible by 6, which is to say 6 times the numbers between 17 and 167. There are thus

$$167 - 17 + 1 = 151$$

of them, and so the correct answer to our problem will be

$$901 - 451 - 300 + 151 = 301.$$ ■

Here's one more involved example of the same idea. Again, we're keeping the convention that by a "word" we mean an arbitrary sequence of letters of the English alphabet.

EXAMPLE 3.3.2 How many four-letter words are there in which no letter appears three or more times in a row?

SOLUTION This clearly calls for the subtraction principle. We know how many four-letter words there are in all—the number is

$$26 \times 26 \times 26 \times 26 = 456{,}976.$$

[1] Is this clear? Think about it, but if you don't see why this is the case, relax; take our word for it now, and in Chapter 8 we'll work it out.

We just have to subtract the number of words in which a letter appears three or more times in a row.

Now, there are two kinds of four-letter words in which a letter appears three times in a row: those in which the first three letters are the same, and those where the last three letters are the same. In each case, the number of such words is easy to count by the multiplication principle. For example, to specify a word in which the first three letters are the same, we have to specify that letter (26 choices) and the last letter (26 choices again), so there are

$$26 \times 26 = 676$$

of this type. By the same token, there are 676 four-letter words in which the last three letters are the same; so naively we want to exclude $2 \times 676 = 1,352$ words.

But once more that's not quite right: the 26 words in which all four letters are the same belong to both classes, and so have been subtracted twice! So to correct the count, we have to add them back in three times. The correct answer is therefore

$$456,976 - 1,352 + 26 = 455,650.$$ ■

Actually, there's another way to do this, morally the same but avoiding the issue of multiple subtractions. We can count the number of words in which one letter appears *exactly* three times in a row, and the number of words in which one letter appears four times, add them up and subtract the total from the number of all four-letter words. For the first, there are again two classes of such words; but within each class the number is different: we choose the repeated letter among the 26 letters of the alphabet as before, but since that letter is to appear exactly three times the remaining letter must be chosen among the remaining 25 letters of the alphabet. There are thus a total of

$$2 \times 26 \times 25 = 1,300$$

such words. There are again 26 words in which one letter appears all four times; so the correct answer is

$$456,976 - 1,300 - 26 = 455,650$$

as before.

This last exercise represents another level of complexity in the subtraction principle, but you should be able to do it if you keep your wits about you.

Exercise 3.3.3 How many five-letter words are there in which no letter appears 3 or more times in a row?

Chapter 4

Collections

In this chapter we're going to introduce a new, fundamental idea in counting. This will also be the last new formula: using this and the ideas we've already introduced in combination, we'll be able to count all the objects we want, at least until the final (and optional) chapter of this part of the book.

There's nothing mysterious about it. Basically, in the last couple of chapters we've considered a range of problems in which we count the number of ways to make a sequence of choices. In each instance, either the choices were made from different collections of objects (shirts and pants; meat and vegetable toppings on our pizza; action thrillers and light romantic comedies) or, if they were selections made from the same collection of objects, the order mattered: when we're counting four-letter words, "POOL" is not the same as "POLO."

What we want to look at now are situations where we choose a collection of objects from the same pool, and *the order doesn't matter*. We'll start by revisiting some of the problems we've dealt with, and show how slight variations will put us in this kind of situation.

4.1 Back to the House of Pizza

It's a new day, and once more you head over to the House of Pizza for lunch. Today, though, you're feeling both hungry and carnivorous: a pizza with three meat toppings sounds about right. Assuming that the HoP is still offering seven meat toppings, how many different pizzas will fit the bill?

On to the video store. Your roommates have once more selected you to go forth and rent the evening's entertainment; this night everyone's in the mood for a festival of movies based on comic strips or video games, and your mission is to return with four of the 23 such movies that the House of Videos carries. How many different choices do you have at the video store?

Finally, back to our class of 15 students. This time, we're not going to select officers; we're just going to choose a committee of four students. There are no distinctions among the four members of the committee; we just have to select four students from among the 15 in the class. How many different committees can be formed?

You get the idea? In each case we're choosing a collection of a specified number of objects from a common pool (toppings, videos, students), and the order doesn't matter: ordering a pizza with sausage, pepperoni and hamburger gets you pretty much the same pizza as ordering one with hamburger, pepperoni and sausage. This sort of situation comes up constantly: when you're dealt a hand of five cards in draw poker, or of 13 cards in bridge, it doesn't matter in what order you receive the cards; possible hands consist of collections of five or 13 cards out of the deck of 52. By way of language, in this sort of situation—where we make a series of selections from a common pool, but all that matters is the totality of objects selected, not the order in which they're selected—we'll refer to choosing a *collection* of objects. In settings where the selections are made from different pools, or where the order does matter, we'll refer to choosing a *sequence*.

Now, as you'll recognize, all of the problems above are really the same problem with different numbers substituted. In fact, there are only two numbers involved: in each of these cases, the number of possible choices depends really only on the number of objects in the pool we're selecting from; and the number of objects to be selected to form our collection.

What we need to do, then, is to find a formula for the number of such collections. We'll do that in the following section, and then we'll see how to combine that formula with the others we've derived to solve a large range of counting problems.

4.2 Binomial Coefficients

The good news: the formula for the number of collections is very simple to write down and to remember. The bad: it's not quite as straightforward to derive as the ones we've done up to now; in fact, figuring it out requires a somewhat indirect argument. What we'll do is show how to find the answer in a particular case, and once we've done that it'll be pretty clear how to replace the particular numbers in that example with arbitrary ones.

Let's take the case of choosing a committee of four students from among a class of 15—that is, the problem of counting the number of possible committees that can be formed. Again, this is a situation where the order of selection doesn't matter: egos aside, choosing Dave and then Rebecca has the same outcome as choosing Rebecca and then Dave; all that matters in the end is who is on the committee and who is not. And since possible committees don't correspond to sequences of choices, the multiplication principle doesn't seem to apply.

But it does apply!—in a sort of weird, backhanded way. To see how, let's focus for a moment on a different problem, the class officer problem: that is, counting the number of ways we can choose a president, vice president, secretary and treasurer for the class, assuming no student can occupy more than one office. As we saw, the multiplication principle works just fine here: we choose a president (15 choices), then the vice president (14), then the secretary (13) and finally the treasurer (12), for total of

$$15 \cdot 14 \cdot 13 \cdot 12 \quad \text{or} \quad \frac{15!}{11!}$$

possible slates.

But now suppose we want to solve the same problem in a different, somewhat warped way (though again using the multiplication principle). Suppose that instead of choosing the slate one officer at a time, we break the process up into two steps: first we choose a committee of four students who will be the class officers, and *then* choose which of those four will be president, vice president, secretary and treasurer.

That may seem like an unnecessarily complicated way to proceed. After all, we already know the answer to the class officer problem, while we don't know the number of committees. Bear with us! Let's look anyway at what it tells us.

The one thing we do know is, having selected the four members of the committee, how many ways there are of assigning to the four of them the jobs of president, vice president, secretary and treasurer: by what we already know, this is just $4 \cdot 3 \cdot 2 \cdot 1 = 4! = 24$. So if we do break up the process of selecting class officers into two stages, choosing a committee and then assigning them the four jobs, what the multiplication principle tells us is that

$$\left\{ \begin{array}{l} \text{the number of ways of} \\ \text{choosing a committee} \end{array} \right\} \cdot 4! = \left\{ \begin{array}{l} \text{the number of ways of} \\ \text{selecting class officers} \end{array} \right\}.$$

Now that, if you think about it a moment, tells us something. Since we know that the number of ways of selecting class officers is $15!/11!$, we can solve this equation for the number of committees:

$$\begin{aligned} \left\{ \begin{array}{l} \text{the number of ways of} \\ \text{choosing a committee} \end{array} \right\} &= \frac{1}{4!} \cdot \left\{ \begin{array}{l} \text{the number of ways of} \\ \text{selecting class officers} \end{array} \right\} \\ &= \frac{15!}{4!11!}. \end{aligned}$$

In English: since every choice of committee corresponds to $4! = 24$ different possible choices of class officers, the number of possible committees is simply $(1/24)^{\text{th}}$ the number of slates.

You can probably see from this that it's going to be the same when we count the number of ways of choosing a collection of any number k of objects from a pool of any number n. We know that the number of ways of choosing a *sequence* of k objects without repetition—a first, then a second different from the first, then a third different from the first two and so on—is just

$$n \cdot (n-1) \cdot \cdots \cdot (n-k+1) = \frac{n!}{(n-k)!}.$$

At the same time, for each possible collection of k objects from the pool, there are

$$k \cdot (k-1) \cdot \cdots \cdot 2 \cdot 1 = k!$$

ways of putting them in order—choosing a first, a second, and so on. The conclusion, then, is that

$$\begin{aligned} &\left\{ \begin{array}{l} \text{the number of ways of choosing} \\ \text{a collection of } k \text{ objects, without} \\ \text{repetition, from a pool of } n \text{ objects} \end{array} \right\} \\ &\qquad = \frac{1}{k!} \cdot \left\{ \begin{array}{l} \text{the number of ways of choosing a} \\ \text{sequence of } k \text{ objects, without} \\ \text{repetition, from a pool of } n \text{ objects} \end{array} \right\} \\ &\qquad = \frac{n!}{k!(n-k)!}. \end{aligned}$$

or, in other words,

> The number of ways of choosing a collection of k objects, without repetition, from among n objects is
>
> $$\frac{n!}{k!(n-k)!}.$$

So, for example, if the House of Pizza offers seven meat toppings, the number of possible pizzas you can order with three meat toppings is

$$\frac{7!}{3!4!} = \frac{5{,}040}{6 \cdot 24} = 35;$$

and if you're sent to the video store with instructions to return with an assortment of exactly four of their 23 movies based on comic strips or video games, your choice is among

$$\frac{23!}{4!19!} = 8{,}855$$

such assortments.

The numbers that appear in this setting are so ubiquitous in math that they have a name and a notation of their own. They're called *binomial coefficients* (for reasons we'll explain in Chapter 6), and written in this way:

$$\binom{n}{k} = \frac{n!}{k!(n-k)!}$$

There are a number of things we can say right off the bat about binomial coefficients. To begin with, there is the basic observation that

$$\binom{n}{k} = \binom{n}{n-k}.$$

This is obvious from the above formula: we see that

$$\frac{n!}{k!(n-k)!} = \frac{n!}{(n-k)!k!}$$

just by rearranging the factors $k!$ and $(n-k)!$ in the denominator. It's also clear from the interpretation of these numbers: after all, specifying which four kids in the class of 15 are to be put on the committee is tantamount to specifying which 11 to leave off it; and in general choosing which k objects to take from a pool of n is the same as choosing which $n-k$ not to take.

Second, as we've pointed out, the standard formula for the binomial coefficients

$$\binom{n}{k} = \frac{n!}{k!(n-k)!}$$

$$= \frac{n \cdot (n-1) \cdot (n-2) \cdots 2 \cdot 1}{k \cdot (k-1) \cdots 2 \cdot 1 \cdot (n-k) \cdot (n-k-1) \cdots 2 \cdot 1}$$

is in some ways not the most efficient way to represent the number—it's certainly not how you would calculate it in practice—since there are factors that appear in both the numerator and the denominator, and can be cancelled. Doing this gives us two alternative ways of writing the binomial coefficient:

$$\binom{n}{k} = \frac{n \cdot (n-1) \cdots (n-k+1)}{k \cdot (k-1) \cdots 2 \cdot 1}$$
$$= \frac{n \cdot (n-1) \cdots (k+1)}{(n-k) \cdot (n-k-1) \cdots 2 \cdot 1}.$$

This is not just an aesthetic issue, it's a practical one as well. Suppose, for example, you wanted to count the number of possible five-card hands from a standard deck of 52—that is, you wanted to evaluate the binomial coefficient $\binom{52}{5}$—and you wanted to carry out the calculation on your calculator. If you write the binomial coefficient as

$$\binom{52}{5} = \frac{52 \cdot 51 \cdot 50 \cdot 49 \cdot 48}{5 \cdot 4 \cdot 3 \cdot 2 \cdot 1}$$

your calculator'll have no trouble multiplying and dividing out the factors. But if you write

$$\binom{52}{5} = \frac{52!}{5!47!}$$

you're in trouble: when you punch in 52! in your calculator you'll probably get an error message; most calculators can't handle numbers that large. Or, even worse, you won't get an error message; your calculator will simply switch to scientific notation. In effect, the calculator will round off the number and not tell you; and these roundoff errors can and often do become significant.

Before we go on, let's look at some special cases of binomial coefficients. To begin with, note that for any n,

$$\binom{n}{1} = n,$$

corresponding to the statement that "there are n ways of choosing one object from among n." (Well, that's not exactly news.) Also, by our convention that $0! = 1$, we see that

$$\binom{n}{0} = \frac{n!}{0!n!} = 1$$

and similarly $\binom{n}{n} = 1$. Again, think of this simply as a convention; it makes the various formulas we're going to discover in Chapter 6 work.

The first interesting case is the number of ways of choosing a pair of objects from among n:

$$\binom{n}{2} = \frac{n(n-1)}{2},$$

so that the number of ways of choosing two objects from among three is $3 \cdot 2/2 = 3$ (remember, this is the same as the number of ways of picking one object); the number of ways of choosing two objects from among four is $4 \cdot 3/2 = 6$; and in general we can make a table

n	*# of ways of choosing two objects from among n*
3	3
4	6
5	10
6	15
7	21
8	28

and so on. We can make a similar table for the binomial coefficients $\binom{n}{3}$:

n	*# of ways of choosing three objects from among n*
4	4
5	10
6	20
7	35
8	56
9	84

Mathematicians have found many fascinating patterns in these numbers, as well as other interpretations of them. We'll take a look at a few of these in Chapter 6.

There's one final remark we want to make about the binomial coefficients, even though it may be of interest only to math nerds. From the formula

$$\binom{n}{k} = \frac{n!}{k!(n-k)!}$$

it's clear that $\binom{n}{k}$ is a fraction, but it's far from clear that it's actually a whole number. Of course, we know it's a whole number from the interpretation as the number of ways of choosing k objects from n, but that just raises the question: can we see *why* the formula above always yields a whole number? In some cases we can do this. For example, when we look at the formula

$$\binom{n}{2} = \frac{n(n-1)}{2}$$

and ask, "why is this a whole number?" we have an answer: no matter what n is, either n or $n - 1$ must be even. Thus the product $n(n - 1)$—the numerator of our fraction—must be even, and so the quotient is a whole number. Likewise, consider the formula

$$\binom{n}{3} = \frac{n(n-1)(n-2)}{6}.$$

Of the three factors n, $n-1$ and $n-2$ that appear in the numerator, at least one must be divisible by 3, and at least one must be even. The numerator must thus be divisible by 6 (again, think about it; if it's not clear now it will be in Chapter 8), and so the quotient is whole.

But it gets less and less obvious as k increases. For example, when we say that

$$\binom{n}{4} = \frac{n(n-1)(n-2)(n-3)}{24}.$$

is a whole number, we're saying in effect that *the product of any four whole numbers in a row is divisible by* 24. Think about this for a moment: could you convince yourself of this fact without using the interpretation of $\binom{n}{4}$? Could you convince someone else?

In fact we'll see in Part III of this book why this is true; for now we'll just leave it as something for you to mull over (or not).

Now it's time for you to do some exercises.

Exercise 4.2.1 Suppose now that the menu at the House of Pizza lists eight meat toppings. How many different pizzas can you order there with two (different) meat toppings? How many with three?

Exercise 4.2.2 Suppose you are given an exam with 10 problems, and are asked to do exactly seven of them. In how many ways can you choose which seven to do?

4.3 Examples

We can combine the formula we have now for the number of collections with other formulas and techniques. Here are some examples, in the form of solved problems.

EXAMPLE 4.3.1 Suppose once more we're asked to choose a collection of four students from our class of 15 to form a committee, but this time we have a restriction: we don't want the committee to consist of all boys or all girls. (Recall that there are eight boys and seven girls in the class.) How many different committees can we form?

SOLUTION This clearly calls for the subtraction principle. We know that the total number of possible committees is

$$\binom{15}{4} = \frac{15 \cdot 14 \cdot 13 \cdot 12}{4 \cdot 3 \cdot 2 \cdot 1} = 1{,}365.$$

The number of committees consisting of all boys is similarly

$$\binom{8}{4} = \frac{8 \cdot 7 \cdot 6 \cdot 5}{4 \cdot 3 \cdot 2 \cdot 1} = 70,$$

and the number of committees consisting of all girls is

$$\binom{7}{4} = \frac{7 \cdot 6 \cdot 5 \cdot 4}{4 \cdot 3 \cdot 2 \cdot 1} = 35.$$

If we exclude those, we see that the number of allowable committees is

$$\binom{15}{4} - \binom{8}{4} - \binom{7}{4} = 1{,}365 - 70 - 35 = 1{,}260.$$

■

Note that even though this is a multiple subtraction, we don't need to add any terms back in, since there are no committees that belong to both excluded classes—that is, a committee can't simultaneously consist of all boys and all girls.

EXAMPLE 4.3.2 One more committee: now suppose we require that the committee includes exactly two boys and two girls. How many possibilities are there now?

SOLUTION This, by contrast, is a case for the multiplication principle: to choose an allowable committee subject to this restriction, we simply have to choose two among the eight boys, and (independently) two among the seven girls. The answer is thus

$$\binom{8}{2} \cdot \binom{7}{2} = 28 \cdot 21 = 588$$ ■

Exercise 4.3.3 Back to the House of Pizza: given that they feature seven meat toppings and four vegetable toppings, how many pizzas can be ordered with two meat and two vegetable toppings?

Exercise 4.3.4 Sam's ice cream shop offers only vanilla ice cream, but has 17 different possible toppings to choose from.

1. How many different sundaes can be formed with *exactly* three toppings?
2. How many different sundaes are there with *at least* two toppings?
3. How many different sundaes can be formed with no restriction on the number of toppings?

EXAMPLE 4.3.5 There are 10 players on a basketball team, and the coach is going to divide them up into two teams of five—the Red team and the Blue team, say—for a practice scrimmage. She's going to do it randomly, meaning that all of the $\binom{10}{5}$ ways of assigning the players to the two teams are equally likely.

Two of the players, Sarah and Rebecca, are friends. "I hope we wind up on the same team," Rebecca says. "Well, we have a 50-50 chance," Sarah replies.

Is Sarah right?

SOLUTION What this problem is asking us to do is to count, among the $\binom{10}{5}$ ways of assigning players to the two teams, how many result in Sarah and Rebecca being on the same team, and how many result in their being on opposing teams. Sarah is, in effect, saying that these two numbers will be equal; we'll calculate both and see if she's right.

Let's start by counting the number of ways of choosing the teams that result in Sarah and Rebecca winding up on the same team. We can specify such an assignment in two stages: first, we decide which team, Red or Blue, gets the Rebecca/Sarah duo. Obviously, there are two possibilities. Having done that, we have to take the remaining eight players and divide them into two groups: three will go to the team that already has Sarah and Rebecca; five will go the other team. The number of ways of doing that is $\binom{8}{3}$, so the total number of team assignments with the two as teammates is

$$2 \cdot \binom{8}{3} = 2 \cdot 56 = 112.$$

Now let's count the number of assignments that result in Sarah and Rebecca opposing each other. Again, we can specify such a choice in two steps: first, we specify which team, Red or Blue, Rebecca is on; Sarah will necessarily go the other. We then have to take the remaining eight players and assign four of them to each of the two teams; there are $\binom{8}{4}$ ways of doing this, and so the number of team assignments with Rebecca and Sarah opposite is

$$2 \cdot \binom{8}{4} = 2 \cdot 70 = 140.$$

The conclusion, then, is that Sarah is wrong: it is more likely that she and Rebecca will wind up on opposing teams.

We're not done! One thing we should always look for in doing these problems is a way to check the accuracy both of our analysis and of our calculations. Here we have a perfect way to do that. We've said that there are a total of $\binom{10}{5}$ ways of assigning the 10 players to the two teams, of which 112 result in the two friends being teammates and 140 result in their being on opposite teams. Before we're satisfied that we've got the correct answer, we should check that in fact $\binom{10}{5}$ is equal to the sum $112 + 140 = 252$. Let's do it:

$$\begin{aligned}\binom{10}{5} &= \frac{10 \cdot 9 \cdot 8 \cdot 7 \cdot 6}{5 \cdot 4 \cdot 3 \cdot 2 \cdot 1} \\ &= \frac{30{,}240}{120} \\ &= 252.\end{aligned}$$

Having done this, we can be much more confident both that our analysis was correct, and that we didn't make any numerical mistakes. We can also say that the probability that Rebecca and Sarah wind up on the same team is

$$\frac{112}{252} = \frac{4}{9} \sim .444$$

or about 44%. We'll talk a lot more about probability in Chapter 5. ■

Exercise 4.3.6 Suppose now that the 10 players are to be divided into teams of six and four. Is it more likely that Sarah and Rebecca will be teammates or opponents?

Note: this problem is just slightly trickier than the one we did above, since in one of the two calculations involved we can't use the multiplication principle. Be sure to check your answer!

EXAMPLE 4.3.7 Let's suppose you're playing Scrabble, and you have in your rack the letters "E, E, E, E, N, N, N" (this sort of thing seems to happen to us a lot). How many ways are there of arranging these tiles in your rack? In other words, how many seven-letter words (once more in the sense of arbitrary strings of letters) are there that contain exactly three Ns and four Es?

SOLUTION This is actually pretty simpleminded, but it'll lead to something more interesting in the next problem. The point is, if we think of our seven-letter word as having seven places to fill with Es and Ns, to specify such a word means just to specify which three of the places are to be filled with Ns; or, equivalently, which four are to be filled with Es. The answer is thus

$$\binom{7}{3} = \binom{7}{4} = 35. \qquad \blacksquare$$

EXAMPLE 4.3.8 This problem may not seem like it has much to do with collections, but we'll see in a moment it does. Suppose we live in a city laid out in a rectangular grid, and that our job is located three blocks north and four blocks east of our apartment, as shown in the picture below.

Clearly, we have to walk seven blocks to get to work each morning. But there are many different paths we could take. How many different paths can we take, if we stay on the grid? Think about it before you look at the answer.

SOLUTION To specify a path from our home to work, we have to give a series of directions like, "Go one block North, then three blocks East, then another block North, then another block East, then another block North," or, for short

N, E, E, E, N, E, N.

In other words, paths correspond exactly to words consisting of exactly 3 Ns and 4 Es. So this is exactly the same problem as the last one and the answer is 35 paths. ■

In general, if we have a k by l rectangular grid, by the same logic the paths from one corner to the opposite one (with no doubling back) correspond to words formed with k Ns (or Ss) and l Es (or Ws). The number of such paths is thus given by the binomial coefficient

$$\binom{k+l}{k},$$

or, equivalently, the binomial coefficient $\binom{k+l}{l}$. Note the symmetry here: the number of paths in a $k \times l$ grid is the same as the number of paths in an $l \times k$ grid, as the formula verifies.

Exercise 4.3.9

1. Consider the grid below. How many paths of shortest possible length (that is, 13 blocks) are there from the point labeled "home" to the point labeled "work"?

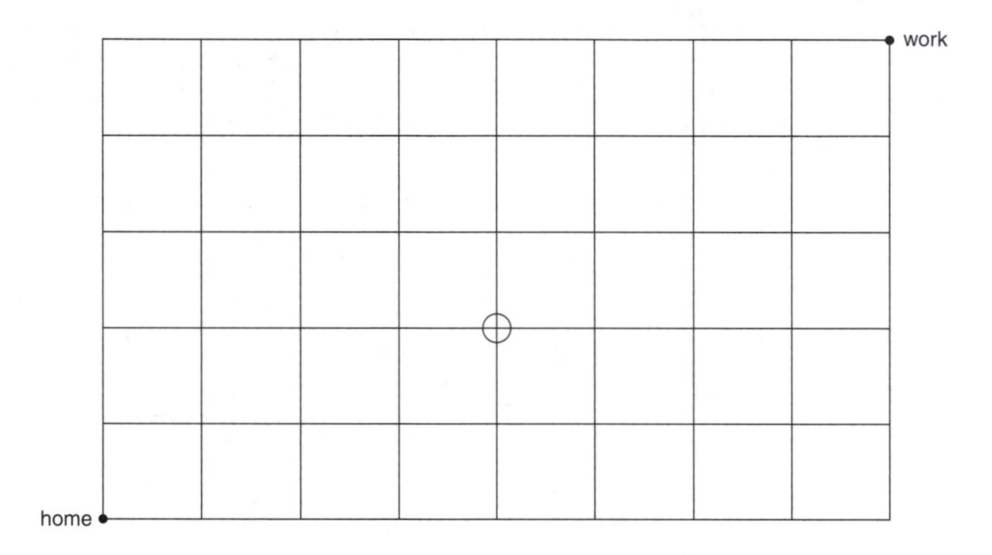

2. Suppose that the small circle on the diagram represents Mike's House of Donuts and Coffee, a crucial stop on the way if you're to arrive at work awake. How many paths from home to work (again of minimal length) pass through Mike's?
3. Suppose by contrast that you're on a strict diet, and that you must at all costs *avoid* passing through the intersection where Mike's is located. Now how many possible paths do you have?

One last item before we conclude this section: we promised in Chapter 2 to show you how to do Example 2.5.2, and now it's time. We'll start by doing a simpler version of the problem that nonetheless introduces the essential ideas.

EXAMPLE 4.3.10 Let's say a class consists of seven boys and seven girls, and we want to line them up so that no two boys are next to each other, and no two girls are next to each other. How many ways are there of doing this?

SOLUTION To start with, the key feature is to separate the problem into two phases:

- First, we have to choose which places in line are to be occupied by boys and which by girls. That is, we have to choose a sequence of genders or, if you like, a 14-letter word consisting of seven Bs and seven Gs; *but with no two Bs adjacent and no two Gs adjacent*. Once we've done that,
- We have to assign an actual student of the appropriate gender to each place.

In Example 2.5.1, the first step didn't exist: since there were eight boys and seven girls, the only possible arrangement of genders was to alternate BGBGBG-BGBGBGBGB. Once we realized that, we simply had to assign the eight boys to the 8 Bs in that sequence, and the seven girls to the seven Gs, for a total of $8! \cdot 7!$ choices.

In the present circumstances, by contrast, there are different possible gender sequences. But not many: since we have to alternate boys and girls, the sequence of genders is determined once we specify the first; that is, it's got to be either GBGBGBGBGBGBGB or BGBGBGBGBGBGBG. So there are just 2 choices for the gender sequence.

The second step is essentially the same in both Example 2.5.1 and the present problem. In the present circumstances, once we've decided on a particular arrangement of genders there'll be exactly 7! ways of assigning the seven boys in class to the seven places in line designated for boys, and 7! ways of assigning the girls' places; so there'll be $7! \cdot 7!$ ways of assigning the 14 students to appropriate places in line. The answer is thus the number of gender arrangements times $7! \cdot 7!$; that is,

$$2 \cdot 7! \cdot 7! = 50{,}803{,}200. \qquad \blacksquare$$

Now we're ready to tackle Example 2.5.2. First, recall the problem: we have a class of 6 boys and 9 girls, and we want to know how many ways to line them up assuming we don't want any two boys next to each other in line. If you haven't thought about the problem, take some time now to do so, especially in light of the example we've just worked out.

Ready? Here goes. The first step, just as in the last problem, is to separate the problem into two phases: specifying which places in line are to be occupied by boys and which by girls; and then assigning an actual student of the appropriate gender to each place. Moreover, the second step is essentially the same in both examples: once we've decided on a particular arrangement of genders there'll be exactly 6! ways of assigning the six boys in class to the six places in line designated for boys, and 9! ways of assigning the girls' places; so there'll be $6! \cdot 9!$ ways of assigning the 15 students to appropriate places in line. The answer is thus $6! \cdot 9!$ times the number of gender arrangements—that is, $6! \cdot 9!$ times the number of 15-letter words consisting of 6 Bs and 9 Gs, with no two Bs in a row.

OK, then, how do we figure out that number? Here is where it gets slightly tricky. The first step is to consider two possibilities: either the sequence ends in a B or it ends in a G. Suppose first that the sequence ends in a G. In that case, we observe, every B in the sequence is necessarily followed by a G. In other words, instead of arranging 6 Bs and 9 Gs—subject to the requirement that no B follow another—we can pair off one G with each B to form six BGs, with three Gs left over, and *count arbitrary arrangements of six BGs with three Gs.* We know how many of those there are; it's just

$$\binom{9}{6} = 84.$$

Next, we consider arrangements ending in a B, and we count those similarly: we take the remaining five Bs and pair each with a G to form five BGs, with four Gs left over. Again, we can take arbitrary arrangements of these five BGs and four Gs, with a B stuck on at the end; and there are

$$\binom{9}{5} = 126$$

of these. Altogether, then, there are $84 + 126 = 210$ possible gender sequences; and since each gives rise to $6! \cdot 9!$ possible lineups, the total number is

$$210 \cdot 6! \cdot 9! = 54{,}867{,}456{,}000.$$

4.4 Multinomials

Say it's now our job to assign college students to dorm rooms. We have a group of nine students to assign, and three rooms to assign them to: one a quad, one a triple and one a double. The standard question: how many different ways can we do it?

"That's nothing new," you might say, "we already know how to do it." And you'd be right: to assign the nine students we could start by choosing four of the nine and assign then to the quad. That leaves five students to be assigned to the triple and double; choose three of the remaining five to go to the triple and you're done, since the remaining two necessarily go in the double. Since we had $\binom{9}{4} = 126$ ways of making the first choice and $\binom{5}{3} = 10$ ways of making the second, the answer would be

$$\binom{9}{4} \cdot \binom{5}{3}.$$

But wait: what would happen if we did the assignments in a different order? Suppose, for example, that we started by assigning two of the students to the double, and then chose four of the remaining seven to go the quad? We'd have $\binom{9}{2}$ choices at the first stage, and $\binom{7}{4}$ at the second, so the correct answer would be

$$\binom{9}{2} \cdot \binom{7}{4}.$$

What's up with that?

The answer is, there's no mistake. When we look closer, we see that

$$\binom{9}{4} \cdot \binom{5}{3} = \frac{9!}{4!5!} \cdot \frac{5!}{3!2!}$$
$$= \frac{9!}{4!3!2!}$$

while on the other hand

$$\binom{9}{2} \cdot \binom{7}{4} = \frac{9!}{2!7!} \cdot \frac{7!}{4!3!}$$
$$= \frac{9!}{4!3!2!}.$$

Hence both approaches lead to the same answer, which is equal to 1,260.

So there's really nothing new here. But the sort of numbers that arise here—the number of ways of distributing some number n of objects into three (or more) collections of specified size—are so common that they, like the binomial coefficients, deserve a name and a notation of their own. Think of it this way: maybe the correct (or at any rate the symmetric) way to think of the binomial coefficient $\binom{n}{k}$ is as the number of ways of distributing a group of n objects into two collections, of size k and $n-k$. Well, in the same vein, whenever we have a number n and three numbers a, b and c that add up to n, we can ask how many ways there are of distributing n objects into three collections, of sizes a, b and c. We can answer this completely analogously to the way we just did the last problem: we first choose which a of our n objects are to go into the first group; then which b of the remaining $n-a$ are to go into the second. That'll leave c objects, which have to go into the third; so by the multiplication principle the number of ways is

$$\begin{aligned}\binom{n}{a}\cdot\binom{n-a}{b} &= \frac{n!}{a!(n-a)!}\cdot\frac{(n-a)!}{b!(n-a-b)!}\\ &= \frac{n!}{a!(n-a)!}\cdot\frac{(n-a)!}{b!c!}\\ &= \frac{n!}{a!b!c!}.\end{aligned}$$

This number is called a *multinomial coefficient*, and is typically denoted by the symbol

$$\binom{n}{a,\ b,\ c} = \frac{n!}{a!b!c!}.$$

Similarly, if a, b, c and d are four numbers adding up to n, the number of ways of distributing n objects into groups of size a, b, c and d is

$$\binom{n}{a,\ b,\ c,\ d} = \frac{n!}{a!b!c!d!},$$

and so on. The most general form of this problem would be: suppose we have n different objects, which we want to distribute into k collections. The number of objects in each collection is specified: the first collection is to have a_1 of the objects, the second a_2 and so on; the kth and last collection is to have a_k of the objects. We ask: how many ways are there of assigning the n objects to the k collections? The answer, as we've suggested, is that

> The number of ways of distributing n objects into groups of size $a_1, a_2, \ldots, a_k$ is
>
> $$\frac{n!}{a_1!\cdot a_2!\cdot\cdots\cdot a_k!}.$$

Again, the number $\frac{n!}{a_1!\cdot a_2!\cdot\ldots\cdot a_k!}$ that appears here is called a multinomial coefficient and denoted $\binom{n}{a_1,a_2,\ldots,a_k}$. Note that in this setting our old friend the binomial coefficient $\binom{n}{k}$ could also be written as $\binom{n}{k,\ n-k}$; but it's easier (and unambiguous) to just drop the $n-k$.

Multinomial coefficients are thus a straightforward generalization of binomial coefficients, and are almost as ubiquitous (though, as we just saw, you don't *really* need to know them: if you just know about binomial coefficients and the multiplication principle, you can solve any problem involving multinomial coefficients).

Exercise 4.4.1 The job has fallen to you to assign 18 incoming freshmen to rooms in one particular dormitory. There are six rooms: two quads, two triples, and two doubles.

1. In how many ways can the 18 freshmen be assigned to the rooms?
2. After you submitted your list of assignments from part (1) to the Dean, she complained that some of them put men and women in the same room. If we designate one of the quads, one of the triples, and one of the doubles for women, in how many ways can the rooms be assigned to nine women and nine men?

One classic example of multinomial coefficients is in counting anagrams. By an *anagram* of a word, we mean a rearrangement of its letters: for example, "SAPS" is an anagram of "PASS." (Note that each letter must appear the same number of times in the anagram as in the original word.) In keeping with our conventions, by an anagram we'll mean an arbitrary rearrangement of the letters, not necessarily a word in the English language.

So: how many anagrams does a word have? In some cases this is easy: if a four-letter word, say, has all different letters (that is, none repeated) then an anagram of the word is simply an ordering of its letters, and so there are 4! of them. For example, the word "STOP" has 24 anagrams

STOP	STPO	SOTP	SOPT	SPTO	SPOT
TSOP	TSPO	TPSO	TPOS	TOSP	TOPS
OSTP	OSPT	OTSP	OTPS	OPST	OPTS
PSTO	PSOT	PTSO	PTOS	POST	POTS

By the same token, an n-letter word with n different letters will have $n!$ anagrams.[1] At the other extreme, the answer is also relatively easy: a word consisting of only one letter repeated n times has no anagrams other than itself; and, as we saw in the example of the Scrabble tiles, if a word consists of k repetitions of one letter and l repetitions of another, it has $\binom{k+l}{k}$ anagrams.

In general, the right way to think about anagrams (from a mathematical point of view) is the way we described in Example 4.3.7. Suppose, for example, we want to count the anagrams of the word "CHEESES." Any such anagram is again a seven-letter word. If we think of it as having seven slots to fill with the letters C, H, E and S then to specify an anagram we have to specify

1. Which one of those seven slots is to be assigned the letter C;
2. Which one of those seven slots is to be assigned the letter H;
3. Which three of those seven slots are to be assigned the letter E; and of course
4. Which two of those seven slots are to be assigned the letter S.

When we think of it in this way, the answer is clear: it's just the multinomial coefficient

$$\binom{7}{1,\ 1,\ 3,\ 2} = \frac{7!}{1!1!3!2!} = 420.$$

[1]For the verbally oriented among you, here's a question. Note that, of the 24 rearrangements of the letters STOP, six of them (STOP, SPOT, OPTS, POST, POTS and TOPS) are actual words in English. Are there four-letter words with more? What five-letter word has the most English word anagrams?

Exercise 4.4.2 How many anagrams does the word "MISSISSIPPI" have? In how many of them are the two Ps next to each other?

Exercise 4.4.3 Consider the following six-letter words: "TOTTER," "TURRET," "RETORT," "PEPPER" and "TSETSE." Which one has the most anagrams, and which the fewest? (You should try and figure out the answer before you actually calculate out the numbers in each case.)

4.5 Something's Missing

At the Bright Horizons School, prizes are given out to the students to reward excellence. All the prizes are identical, though the school may choose to give more than one prize to a given student.

In Ms. Wickersham's class at B.H.S., there are 14 students: Alicia, Barton, Carolina, and so on up to Mark and Nancy. Ms. W. has eight prizes to award, and has to decide how to give them out—that is, how many prizes each child should get. In how many ways can she do this?

For a slightly different formulation, suppose for the moment that you're the chief distributor for the National Widget Importing Co. The NWI has 14 warehouses, called (the NWI is not a very fanciful outfit) Warehouse A, Warehouse B and so on up to Warehouse N.

One day, eight containers of widgets show up at the docks, and it's your job to say how many of the eight should go to each of the 14 warehouses. How many ways are there of doing this?

Or: you're in the dining hall one day, and there's a massive fruit bowl, featuring unlimited quantities of each of 14 different fruits: apples, bananas, cherries and so on up to nectarines. Feeling a mite peckish, you decide to help yourself to eight servings of fruit, possibly taking more than one serving of a given kind. How many different assortments can you select?

Well, what is the point here? Actually, there are a couple: one, we don't know how to solve this problem; and two, we should. Think about it: we've derived, so far in this book, three formulas for counting the number of ways of making a series of k selections from a pool of n objects:

- We know the number of ways of choosing a sequence (that is, the order does matter) from the pool, with repetitions allowed: it's n^k.
- We know the number of ways of choosing a sequence from the pool, with no repetitions: it's $n!/(n-k)!$.
- We know the number of ways of choosing a collection (that is, the order doesn't matter) from the pool, with no repetitions: it's $\binom{n}{k} = \frac{n!}{k!(n-k)!}$.

If we arrange these formulas in a table, as here:

	repetitions allowed	*without repetitions*
sequences	n^k	$\dfrac{n!}{(n-k)!}$
collections	??	$\dfrac{n!}{k!(n-k)!}$

it's clear that something's missing: we don't have a formula for the number of collections of k objects, chosen from a pool of n, with repetitions allowed. That's what all those problems we just listed (or that single problem we repeated three times) involve.

So: are we going to tell you the answer, already? Well, yes and no. We are going to work out the formula in Chapter 7, at the end of this part of the book. But we thought it'd be nice to leave you something to think about and work on in the meantime. So we'll leave it as a challenge: can you solve the problem(s) above before we get to Chapter 7?

YOU FORGOT TO ADD THE VECTORS.
LOOK, LET'S JUST FORGET IT.
WHAT DO YOU MEAN, "FORGET IT"?! I THOUGHT YOU WANTED TO LEARN THIS STUFF!
I DO. BUT IT'S OBVIOUS THAT I CAN'T. MY BRAIN ISN'T BUILT FOR MATH. THAT'S ALL THERE IS TO IT. IT'S JUST TOO HARD!
AMEND
YOU CAN'T QUIT JUST BECAUSE SOMETHING IS "HARD"! YOU HAVE TO STICK WITH IT! YOU HAVE TO TOUGH IT OUT! I WON'T LET YOU QUIT! I WON'T! I WON'T!
I'LL STILL PAY YOU THE FULL $5.
TOODLES.
I'VE LEARNED SOMETHING.

Chapter 5

Probability

In this chapter we're going to use our counting skills to discuss some basic problems in probability. We'll focus primarily on games—flipping coins, rolling dice, and playing poker and bridge—but it should be clear how the same ideas can be applied in other areas as well.

5.1 Flipping Coins

Suppose we flip a coin six times. What's the probability of getting three heads and three tails?

To answer this, we have to start with two hypotheses. The first is simply that we have a fair coin—that is, one that on average will come up heads half the time and tails half the time.

To express the second hypothesis, we have to introduce one bit of terminology. By the *outcome* of the process of flipping the coin six times we mean the sequence of six results, which we can think of as a six-letter word consisting of Hs and Ts. How many possible outcomes are there? That's easy: by the very first formula we worked out using the multiplication principle, the number of such sequences is 2^6, or 64.

Now, it is a fundamental hypothesis of probability that *all 64 outcomes are equally likely*. In effect, this means simply that the result of each coin flip is equally likely to be heads or tails, irrespective of what the result of the previous flips might have been—in English, "the coin has no memory." We should emphasize here that this is really a hypothesis: even though we're all brought up nowadays to see this as self-evident, and it's been verified extensively by experiment, it's not something we can logically prove. Indeed, there were long periods of human history when just the opposite was thought to be true: when people (and not just degenerate gamblers) believed, for example, that after a long run of heads a tail was more likely than another head.

So: let's adopt these hypotheses. What they mean is that any specific outcome—three heads followed by three tails (HHHTTT), or three tails followed by three heads (TTTHHH), or whatever—will occur $\frac{1}{64}$ of the time; in other words, the probability is 1 in 64 that any specified outcome will occur on a given experiment of six flips. Given this, to determine the likelihood of getting exactly three heads and three tails on a given six flips, we have to answer a counting problem: of the 64 possible outcomes, how many include three Hs and three Ts?

This is also an easy problem: the number of six-letter words consisting of three Hs and three Ts is just the binomial coefficient

$$\binom{6}{3} = 20.$$

Now, if each of these outcomes occurs $\frac{1}{64}$ of the time, then in the aggregate we would expect one of these 20 outcomes to occur

$$\frac{20}{64} = \frac{5}{16} = .3125$$

of the time. In other words, when we flip six coins, we expect to get an equal number of heads and tails a little less than one-third of the time.

You can probably see the general rule here: if we flip a coin n times, there are 2^n possible outcomes—corresponding to the n-letter words consisting entirely of Hs and Ts—each of which will on average occur one in 2^n times. The number of these outcomes that involve exactly k heads and $n-k$ tails is $\binom{n}{k}$; and so our conclusion is that

> The probability of getting exactly k heads in n flips is $\dfrac{\binom{n}{k}}{2^n}$.

We'll look further into other probabilities associated with flipping coins, but before we do we should take a moment to point out that this is the basic paradigm of probability: in general, when all possible outcomes of an experiment or process are equally likely, and we separate out the collection of outcomes into two kinds, favorable and unfavorable, the probability of a favorable outcome is simply

$$\text{probability of a favorable outcome} = \frac{\text{number of favorable outcomes}}{\text{total number of possible outcomes}}.$$

Note again that this presupposes that all outcomes are equally likely. If that's not the case, or if we define "outcome" incorrectly, all bets are off, so to speak.

Here are some more examples. As with all probability problems, it's fun to think about them a little and try to estimate the odds before you actually go ahead and calculate them: sometimes they can surprise you (and you can come up with some lucrative bets).

EXAMPLE 5.1.1 Let's say you flip a coin eight times. What is the probability of getting three or more heads?

SOLUTION We have to figure out, of the $2^8 = 256$ eight-letter words consisting entirely of Hs and Ts, how many have at least three Hs. It's slightly easier to figure out how many don't, and use the subtraction principle: we have to count the number of such words that have zero, one or two Hs, and by what we've done the number is

$$\binom{8}{0} + \binom{8}{1} + \binom{8}{2} = 1 + 8 + 28 = 37.$$

The number of such sequences that do have three or more heads is thus $256 - 37 = 219$, and so the probability of getting at least three heads is

$$\frac{219}{256} \sim .85.$$

In other words, you'll get three or more heads in eight flips about 85% of the time. ■

EXAMPLE 5.1.2 Say you and a friend are gambling. You flip nine coins; if they split 4/5—that is, if they come up either four heads and five tails or four tails and five heads—you pay him \$1; otherwise, he pays you \$1. Who has the better odds?

SOLUTION There are $2^9 = 512$ possible outcomes of the nine coin flips, of which

$$\binom{9}{4} + \binom{9}{5} = 126 + 126 = 252$$

involve either four or five heads. That leaves

$$512 - 252 = 260$$

outcomes that don't. Thus the odds are (very slightly) in your favor. ■

EXAMPLE 5.1.3 A variant of the last problem. You and your friend flip six coins; if three or more come up heads you pay him \$1; if two or fewer are heads, he pays you \$2. Who has the better odds?

SOLUTION This is more complicated than the last problem only in that the payoffs are different. We start the same way: figuring out, of the $2^6 = 64$ possible outcomes of the six coin flips, how many result in a win for you and how many in a win for your friend. First, the number of outcomes with fewer than three heads—that is, with 0, 1 or 2 heads—is the sum

$$\binom{6}{0} + \binom{6}{1} + \binom{6}{2} = 1 + 6 + 15 = 22.$$

On these 22 outcomes, you win \$2. That leaves

$$64 - 22 = 42$$

outcomes where your friend wins \$1. You'll lose \$1, in other words, slightly less than twice as often as you win \$2; so once again the odds are slightly in your favor. ■

Exercise 5.1.4 You and a friend play the following game. You each flip three coins, and whoever gets more heads wins; if you get the same number, you win. If you win, your friend pays you \$1; if your friend wins, you pay her \$2. Who has the better odds?

Exercise 5.1.5 Say you flip a coin five times. What's the probability that some three in a row will come up the same?

5.2 Tumbling Dice

As far as mathematics goes, dice are not that different from coins; this section won't introduce any new ideas. But because dice have six faces rather than two, and you can do things like add the results of several dice rolls, they're more interesting. (Las Vegas casinos have tables for playing craps; they don't have coin-flipping tables.)

Let's start by rolling two dice and calculating some simple odds. Again, the hypotheses: first, the dice are fair; in other words, each of the six faces will on average come up one-sixth of the time. Second, if we define the outcome of the roll to be a sequence of two numbers between 1 and 6, the $6 \times 6 = 36$ possible outcomes are all equally likely, that is, each occurs $\frac{1}{36}$ of the time.

We should stop for a moment here and try to clarify one potential misunderstanding. Most of the time, when we roll a pair of dice, the two dice are indistinguishable and we don't think of one as "the first die" and the other as "the second die." *But for the purposes of calculating odds, we should.* For example, there are two ways of rolling "a 3 and a 4": the first die could come up 3 and the second 4, or vice versa; "a 3 and a 4" thus comes up $\frac{2}{36}$, or $\frac{1}{18}$, of the time. By contrast, "two 3s" arises in only one way, and so occurs only $\frac{1}{36}$ of the time. This can be confusing, and even counter-intuitive: when we roll two identical dice, we may not even know whether we've rolled "a 3 and a 4" or "a 4 and a 3." It may help to think of the dice as having different colors—one red and one blue, say—or of rolling them one at a time, rather than together.

With this said, let's calculate some odds. To begin with, let's say we roll two dice and add the numbers showing. We could ask, for example: what's the probability of rolling a 7?

To answer that, we simply have to figure out, of the 36 possible outcomes of the roll, how many yield a sum of 7? This we can figure out by hand: we could get a 1 and a 6, a 2 and a 5, a 3 and a 4, a 4 and a 3, a 5 and a 2 or a 6 and a 1, for a total of six outcomes. The probability of rolling a 7, accordingly, is $\frac{6}{36}$, or $\frac{1}{6}$.

By contrast, there is only one way of rolling a 2—both dice have to come up 1—so that'll come up only $\frac{1}{36}$ of the time. Similarly, there are two ways of rolling a 3—a 1 and a 2, or a 2 and a 1—so that arises $\frac{2}{36}$, or $\frac{1}{18}$ of the time. You can likewise figure out of the odds of any roll; you should take a moment and verify the probabilities in the table below.

sum	*# of ways to achieve the sum*	*probability*
2	1	1 in 36
3	2	1 in 18
4	3	1 in 12
5	4	1 in 9
6	5	5 in 36
7	6	1 in 6
8	5	5 in 36
9	4	1 in 9
10	3	1 in 12
11	2	1 in 18
12	1	1 in 36

Now let's look at some examples involving three or more dice.

EXAMPLE 5.2.1 Suppose now you roll three dice. What are the odds that the sum of the faces showing will be 10? What are the odds of rolling a 12? Which is more likely?

SOLUTION There are 6^3, or 216, possible outcomes of the roll of three dice; we just have to figure out how many add up to 10, and how many add up to 12.

There are many ways of approaching this problem—we could even just write out all the possible outcomes, but it's probably better to be systematic. Here's one way: if the sum of the first two rolls is any number between 4 and 9, then there is one and only one roll of the third die that will make the sum of all three equal to 10. Thus, the number of ways we can get 10 is simply the sum of the number of outcomes of two dice rolls that add up to 4, the number of outcomes of two dice rolls that add up to 5 and so on up to the number of outcomes of two dice rolls that add up to 9. We worked all these out a moment ago; the answer is

$$3 + 4 + 5 + 6 + 5 + 4 = 27.$$

Thus the probability of rolling a 10 with three dice is 27 out of 216, or simply 1 in eight.

Similarly, the number of ways we can get 12 is simply the sum of the number of outcomes of two dice rolls that add up to 6, the number of outcomes of two dice rolls that add up to 7 and so on up to the number of outcomes of two dice rolls that add up to 11; that is,

$$5 + 6 + 5 + 4 + 3 + 2 = 25.$$

So the probability of rolling a 12 with three dice is slightly less than the probability of rolling a 10. ■

EXAMPLE 5.2.2 Let's again roll three dice; this time, calculate the probability of getting at least one 6.

SOLUTION This is actually simpler than the last problem, because it's easier to be systematic. Just use the subtraction principle: the number of outcomes that include at least one 6 is 216 minus the number of outcomes that don't involve a 6; that is,

$$216 - 5^3 = 216 - 125 = 91.$$

The probability of getting at least one 6 on three rolls is thus 91 out of 216. ■

EXAMPLE 5.2.3 For a final example, let's roll seven dice. What is the probability of getting exactly two 6s?

SOLUTION This time, it's the multiplication principle we want to use. We know there are $6^7 = 279{,}936$ sequences of seven numbers from 1 to 6; we have to count how many such sequences contain exactly two 6's. Well, we can specify such a sequence by choosing, in turn:

- Which two of the seven numbers in the sequence are to be the 6's; and
- What the other five numbers in the sequence are to be.

For the first, the number of choices is just $\binom{7}{2}$, or 21. The second involves simply specifying a sequence of five numbers other than 6, that is, from 1 to 5; the number of choices is thus $5^5 = 3{,}125$. The total number of the sequences we're counting is thus

$$21 \times 3{,}125 = 65{,}625$$

and the probability of rolling such a sequence is

$$\frac{65{,}625}{279{,}936} \sim .234.$$

In other words, our chances are slightly less than 1 in 4. ■

Exercise 5.2.4 Say you roll five dice. What are the odds that you'll get at least one 5 and at least one 6?

Exercise 5.2.5 In the game of Phigh, each player rolls three dice; his or her score is the highest number that appears.

1. What is the probability of scoring 1?
2. What is the probability of scoring 2?
3. Your opponent scored 4. What is the probability that you'll win (that is, score 5 or 6)?

5.3 Playing Poker

It's time to graduate from dice to cards, and we're going to focus here primarily on probabilities associated with poker.

To start with, let's establish the rules. A standard deck consists of 52 cards. There are four suits: spades (♠), hearts (♡), diamonds (◇) and clubs (♣). There are 13 cards of each suit, with denominations 2, 3, 4 up to 10, jack (J), queen (Q), king (K) and ace (A). A poker hand consists of five cards; the ranks of the various hands are as follows:

A pair: a hand including two cards of the same denomination
Two pair: a hand including two cards each of two denominations
Three of a kind: a hand including three cards of the same denomination
Straight: a hand in which the denominations of the five cards form an unbroken sequence. For this purpose an ace may be either high or low; that is, A 2 3 4 5 and 10 J Q K A are both straights.
Flush: a hand in which all five cards belong to the same suit
Full house: a hand consisting of three cards of one denomination and two cards of another denomination
Four of a kind: a hand including four cards of the same denomination
Straight flush: a hand consisting of five cards of the same suit forming an unbroken sequence

Note that when we talk about a hand whose rank is "exactly three of a kind," we'll mean a hand of that rank and no higher.

We're going to start with the basic question: if you're dealt five cards at random, what are the odds of getting a given type of hand? Here "at random" means that all the possible hands, of which there are

$$\binom{52}{5} = \frac{52 \cdot 51 \cdot 50 \cdot 49 \cdot 48}{5 \cdot 4 \cdot 3 \cdot 2 \cdot 1} = 2{,}598{,}960,$$

are equally likely to arise; so that the odds of getting a particular type of hand are just the total number of such hands divided by 2,598,960. Our goal, then, will be (as usual) to count the number of hands of each type.

We'll start at the top, with straight flushes. These are straightforward to count via the multiplication principle: to specify a particular straight flush, we simply have to specify the denominations and the suit, which are independent choices. There are four suits, obviously; and as for the denominations, a straight can have as its low card any card from A up to 10 (remember that we count A 2 3 4 5 as a straight), so there are 10 possible denominations. There are thus

$$4 \times 10 = 40$$

straight flushes, and the probability of being dealt one in five cards is accordingly

$$\frac{40}{2{,}598{,}960} \sim .0000153$$

or approximately 1 in 64,974. Not an everyday occurrence: if you play, for example, on the order of two hundred hands a week, it'll happen to you roughly once in six or seven years.

Next is four of a kind. Again, the multiplication principle applies more or less directly: to specify a hand with four of a kind, we have to specify first the denomination of the four, and then say which of the remaining 48 cards in the deck will be the fifth card of the hand. The number of choices is accordingly

$$13 \times 48 = 624,$$

and the probability of being dealt one in five cards is accordingly

$$\frac{624}{2{,}598{,}960} \sim .00024001$$

or, in cruder terms, approximately 1 in 4,000. A good bit more likely than a straight flush, in other words, but don't hold your breath; again, if you play on the order of two hundred hands a week, on average you'll get two or three of these a year.

Note that if we wanted to calculate the odds of getting "four of a kind or better" we'd have to add the number of hands with four of a kind and the number of hands with a straight flush. In general, we're going to calculate here the odds of getting a hand of exactly a given rank; to count the number of hands of a specified rank *or higher* you'll have to add up the numbers of hands of each rank above.

Full houses are also straightforward to count. Since a full house consists of three cards of one denomination and two cards of another, we have to specify first the denomination of which we have three cards, and the denomination of which we have two; then we have to specify which three of the four cards of the first denomination are in the hand, and which two of the second. Altogether, then, the number of choices is

$$\begin{aligned} 13 \times 12 \times \binom{4}{3} \times \binom{4}{2} &= 13 \times 12 \times 4 \times 6 \\ &= 3{,}744. \end{aligned}$$

The probability is

$$\frac{3{,}744}{2{,}598{,}960} \sim .0014406$$

or approximately 1 in 700.

Flushes are even easier to count: we specify a suit (4 choices), and then which five of the 13 cards in that suit will constitute the hand. The number of flushes is thus

$$4 \times \binom{13}{5} = 4 \times 1{,}287$$
$$= 5{,}148.$$

Remember, though, that this includes straight flushes too! If we want to count the number of hands with *exactly* a flush and not any higher rank, we have to subtract those 40 hands, so that the number is

$$5{,}148 - 40 = 5{,}108.$$

The probability is thus

$$\frac{5{,}108}{2{,}598{,}960} \sim .0019654$$

or approximately 1 in 500.

If you're with us so far, straights are likewise simple to count: we have to specify the denominations of the cards—10 choices, as we counted a moment ago—and then the suits, which involve four choices for each of the five cards. The total number of straights is therefore

$$10 \times 4^5 = 10 \times 1{,}024$$
$$= 10{,}240;$$

and if we exclude straight flushes, the number of hands whose rank is exactly a straight is

$$10{,}240 - 40 = 10{,}200.$$

The probability is

$$\frac{10{,}200}{2{,}598{,}960} \sim .0039246$$

or very approximately 1 in 250.

Next, we count hands with exactly three of a kind. Initially, this is similar to the cases we've done before: we have to specify the denomination of the three (13 choices); which three of the four cards of that denomination are to be in the hand $\left(\binom{4}{3} = 4 \text{ choices}\right)$, and finally which two of the remaining 48 cards of the deck will round out the hand.

But here there's one additional wrinkle: since we're counting only hands with the rank "three of a kind," and not "full house," the last two cards can't be of the same denomination. Now, if we were counting the number of sequences of two cards of different denominations from among those 48, the answer would be immediate: we have 48 choices for the first, and 44 for the second, for a total of $48 \times 44 = 2{,}112$ choices. Since the order doesn't matter, though, and because each collection of two such cards corresponds to two different sequences, the number of pairs of cards of different denominations from among those 48 is

$$\frac{48 \cdot 44}{2} = 1{,}056.$$

The number of hands with exactly three of a kind is thus

$$13 \times 4 \times 1{,}056 = 54{,}912$$

and the probability is

$$\frac{54{,}912}{2{,}598{,}960} \sim .021128$$

or roughly 1 in 50. In other words, if your typical night of poker consists of two hundred hands, you're likely to be dealt three of a kind four times.

Counting hands with exactly two pair is slightly easier. We specify the two denominations involved; we have $\binom{13}{2} = 78$ choices there. Then we have to say which two of the four cards of each of these denominations go in the hand; that's

$$\binom{4}{2}^2 = 6^2 = 36$$

choices. Finally, we have to say which of the remaining 44 cards in the deck (of the other 11 denominations) will complete the hand. The total number is, accordingly,

$$78 \times 36 \times 44 = 123{,}522$$

and the probability is

$$\frac{123{,}522}{2{,}598{,}960} \sim .047539$$

or approximately 1 in 20.

Finally, we come to the hands with exactly one pair. We can do this in a similar fashion to our count of hands with three of a kind; choose the denomination of the pair (13); choose two cards of that denomination $\left(\binom{4}{2} = 6\right)$, and finally choose three cards among the 48 cards not of that denomination. But, as in the case of hands with three of a kind, there's a wrinkle in that last step: the three cards not part of the pair must all be of different denominations. Again, if we were counting sequences, rather than collections, of cards, this would be straightforward: there'd be $48 \times 44 \times 40$ choices. Since we have to count collections, however, and since every collection of three cards corresponds to $3! = 6$ different sequences, the number of such collections is

$$\frac{48 \cdot 44 \cdot 40}{6} = 14{,}080.$$

The number of hands with exactly a pair is thus

$$13 \times 6 \times 14{,}080 = 1{,}098{,}240$$

and the probability is

$$\frac{1{,}098{,}240}{2{,}598{,}960} \sim .42256$$

or a little worse than half. As we remarked before, if we want to find the odds of being dealt a pair *or better*, we have to add up the numbers of all hands better than a pair: the total is

$$40 + 624 + 3{,}744 + 5{,}108 + 10{,}200 + 54{,}912 + 123{,}552$$
$$+ 1{,}098{,}240 = 1{,}296{,}420.$$

Now, there's another way to calculate this number, and it gives us a way to check a lot of our calculations. We can count the number of hands with a pair or better by the subtraction principle: that is, count the hands with no pairs, and subtract that from the total number of hands. To count the hands with no two cards of the same denomination, as in the calculation we did of hands with three of a kind or with a pair, we count first the sequences of five cards, no two of the same denomination; this number is simply

$$52 \times 48 \times 44 \times 40 \times 36.$$

But every such hand of cards corresponds to $5! = 120$ sequences, so the number of hands with no two cards of the same denomination is

$$\frac{52 \times 48 \times 44 \times 40 \times 36}{120} = 1{,}317{,}888.$$

But we're not quite done: these 1,317,888 hands include straights, flushes and straight flushes, and if we want to count hands that *rank below* a pair, we have also to exclude these. Thus the total number of poker hands ranking below a pair will be

$$1{,}317{,}888 - 40 - 5{,}108 - 10{,}200 = 1{,}302{,}540,$$

and the number of hands ranked a pair or better will be

$$2{,}598{,}960 - 1{,}302{,}540 = 1{,}296{,}420$$

as we predicted. Note that the probability of getting a pair or better is thus

$$\frac{1{,}296{,}420}{2{,}598{,}960} \sim .49843$$

or very nearly one in two.

Exercise 5.3.1 What are the odds of being dealt a *busted flush*—that is, four cards of one suit and a fifth card of a different suit?

Exercise 5.3.2 What are the odds that a five-card poker hand will contain at least one ace?

Exercise 5.3.3 This doesn't actually have anything to do with poker,[1] but what are the odds that a five-card poker hand will consist entirely of cards of the same color? Is this the same as the odds that 5 coins flipped will all come up the same? Why, or why not?

5.4 Really Playing Poker

This section is probably unnecessary, but our lawyers insisted that we include it.

The odds we've just calculated are obviously relevant to playing poker, but they're only the tip of the tip of the iceberg. In almost all versions of poker, your hand isn't simply dealt to you all at once; it comes in stages, after each of which there's a round of betting. Each time you have to bet, you have to calculate the

[1]Unless, of course, you're playing with extremely nearsighted people and can pass one of these off as a flush.

likelihood of winding up with each possible hand, based on what you have already and how many cards you have yet to receive.

What's more, most poker games involve at least some cards dealt face up, and every time a card is dealt face up, it changes the odds of what you're likely to receive on succeeding rounds, and what your hand is likely to wind up being. In addition, every time someone bets (or doesn't) or raises (or doesn't), it changes the (estimated) odds of what their hole cards are, and hence what cards you're apt to be dealt on succeeding rounds. In fact, every time it's your bet, you have to calculate the odds of your achieving each possible hand, and the amount you stand to win or lose depending on what you get (which depends in turn on other factors: what the other players get, how much is currently in the pot, how much the other players will contribute to the pot and how much you'll have to contribute to the pot).

To be really good at poker, you have to be able to calculate these odds accurately (and unsentimentally). At the same time, it can never be exact: for one thing, no one can make that many calculations that quickly. For another, figuring out how likely it is that the player across the table really does have a king under is necessarily an inexact science. In other words, serious poker exists somewhere in that gray area between mathematics and intuition. Those of us with weaknesses in either field should probably limit our bets.

5.5 Bridge

Bridge is a card game that calls, as much as poker, for estimations of odds. We're not going to discuss the game in any depth or detail at all, but there is one aspect of the game that makes for a beautiful problem in probability, which we'll describe.

In bridge, each player is dealt a hand of 13 cards from a standard deck. This means there are

$$\binom{52}{13} = 635{,}013{,}559{,}600$$

possible hands, all of which we're going to assume are equally likely on any given deal. Now, every hand has what is called a *distribution*, meaning how many cards it has from the four different suits: for example, a hand with four cards of one suit and three each of the others is said to have a 4333 distribution; a hand with four cards each of two suits, three of a third and two of the final suit is said to have a 4432 distribution, and so on.

The question we want to take up here is: what are the odds of a bridge hand having a given distribution? As a special case, we could ask: which is more likely to occur, a 4333 distribution, or a 4432? How do the odds of either compare to the odds of getting a relatively unbalanced distribution, like 5431?

Let's start by counting the number of hands with a 4333 distribution. Basically, we can specify such a hand in two stages: first, we specify which suit is to be the four-card suit; and then we have to specify which of the 13 cards from each suit we're to receive. To specify the four-card suit, there are clearly 4 choices; and as for specifying which of the 13 cards from each suit we're to receive, we have to choose four cards from one suit and three from each of the others. By the multiplication principle, then, the total number of choices is

$$4 \cdot \binom{13}{4} \cdot \binom{13}{3} \cdot \binom{13}{3} \cdot \binom{13}{3}$$

or, in factorials,

$$= 4 \cdot \frac{13!}{4!9!} \cdot \frac{13!}{3!10!} \cdot \frac{13!}{3!10!} \cdot \frac{13!}{3!10!}$$

which works out to

$$= 66{,}905{,}856{,}160.$$

The probability of being dealt a hand with a 4333 distribution is thus

$$\frac{66{,}905{,}856{,}160}{635{,}013{,}559{,}600} \sim .105,$$

or slightly better than one in ten.

Let's do the 4432 distribution next. The idea is the same: first we figure out how many ways we can match the four numbers with the four suits; then, once we've specified how many cards of each suit we're to receive, we calculate how many ways we can choose those cards. For the first part, we have to choose the suit with two cards (four choices) and then the suit with three (three choices); the remaining two suits will each get four cards. The total number of choices is thus

$$4 \cdot 3 \cdot \binom{13}{4} \cdot \binom{13}{4} \cdot \binom{13}{3} \cdot \binom{13}{2}$$

$$= 4 \cdot 3 \cdot \frac{13!}{4!9!} \cdot \frac{13!}{4!9!} \cdot \frac{13!}{3!10!} \cdot \frac{13!}{2!11!}$$

$$= 136{,}852{,}887{,}600.$$

The probability of being dealt a hand with a 4432 distribution is thus

$$\frac{136{,}852{,}887{,}600}{635{,}013{,}559{,}600} \sim .216,$$

or slightly better than one in five. So in fact we see that you're more than twice as likely to be dealt a hand with a 4432 distributions as one with a 4333 distribution!

By now you've probably got the idea; so you can do some yourself:

Exercise 5.5.1 Calculate the probability of being dealt a hand with

1. a 5332 distribution;
2. a 4441 distribution; and
3. a 7321 distribution.

Before you make the calculations, guess which will be most likely and which least.

The next exercise has to do with a basic problem in bridge: once you've seen your cards, what are the odds that govern what everyone else's hand looks like? Clearly what you've got has some effect on the odds: if you have 11 spades, for example, you can be certain that no one at the table has a 4333 distribution. It's a hard problem, but if you can do it you can call yourself a master counter.

Exercise 5.5.2 Say you're playing bridge, and you pick up your hand to discover you have a 7321 distribution. What are the odds that the player to your left has a 4333 distribution?

5.6 The Birthday Problem

Everyone has a birthday; and, leaving aside for the moment those unfortunate souls born on February 29 of a leap year, everyone's birthday is one of the 365 days of the standard year. The probability of two people selected at random having the same birthday is, accordingly, 1 in 365.

So, suppose now we get 10 people together at random. What are the odds that two have the same birthday? How about a group of 25, or 50, or 100? It's pretty clear that as the number of people in the group increases, so does the probability of two people having the same birthday—when you get up to 366 people, of course, it's a lock—so we might ask: *for what size group is there actually a better than 50% chance of two people having the same birthday*? We know how to calculate the odds by now, of course, but before we do so you might want to take a few moments out and think about it—take a guess.

Time's up; here we go. Suppose we line up a group of, say, 50 people, and list their birthdays. We get a sequence of 50 days of the year; and assuming the people were picked randomly—so that each one is as likely to have been born on one day as another—of the 365^{50} possible such sequences, all are equally likely.

So: how many of these 365^{50} possible sequences involve a repeated day? Well, we know how many don't: the number of sequences of 50 days without repetition is, by the standard formula, the product

$$\frac{(365)!}{(315)!} \qquad \text{or} \qquad 365 \cdot 364 \cdot 363 \cdot \cdots \cdot 317 \cdot 316.$$

The probability of there *not* being a repeated birthday among 50 people is thus

$$\frac{365 \cdot 364 \cdot 363 \cdot \cdots \cdot 317 \cdot 316}{365^{50}}.$$

Now, these are some hefty numbers, and we have to be careful how we multiply them out: if we just ask our calculator to come up with 365^{50}, it'll never speak to us again. But we can rewrite this in a form that keeps the numbers reasonably sized:

$$\frac{365 \cdot 364 \cdot 363 \cdot \cdots \cdot 317 \cdot 316}{365^{50}} = \frac{365}{365} \cdot \frac{364}{365} \cdot \cdots \cdot \frac{317}{365} \cdot \frac{316}{365}$$

$$= 1 \cdot \left(1 - \frac{1}{365}\right) \cdot \left(1 - \frac{2}{365}\right) \cdot \cdots \cdot \left(1 - \frac{48}{365}\right) \cdot \left(1 - \frac{49}{365}\right).$$

This is something we (or rather our computers) can evaluate, and the answer is that the probability of not having a repeat is 0.0296. In other words, if we take 50 people at random, the probability is better than 97% that two will share a birthday! Pretty surprising, when you think about it.

In fact, if you work it out, it's already the case with 23 people that the probability of a repeated birthday is 50.7%, or better than half; and by the time you get to 30 people the probability is 70.6% that two people will have the same birthday.

Chapter 6

Pascal's Triangle and the Binomial Theorem

If you're with us so far—if most of the calculations in the last chapter make sense to you—you've got a pretty good idea of what counting is about. In particular, you've seen all the ideas and techniques of counting that we're going to use in the rest of this book. From a strictly logical point of view, you could proceed directly to the second part.

But in the course of our counting, we've come across a class of numbers, the binomial coefficients, that are worth studying in their own right, both for the fascinating properties and patterns they possess and for the way they crop up in so many areas of mathematics. We're going to take some time out here, accordingly, and devote a chapter to the binomial coefficients themselves. These detours are common in mathematics—the tools that we develop to solve a particular problem often open up surprising areas of investigation in their own right.

6.1 Pascal's Triangle

Probably the best way to go about looking for patterns in binomial coefficients is simply to make a table of them and stare at it—maybe we'll be able to deduce something. (Mathematicians like to give the impression that they arrive at their conclusions by abstract thought, but the reality is more prosaic: most of us at least start with experimentation.) As for the form this table should take, there's a classic way of representing the binomial coefficients that is particularly well-suited to displaying their patterns, called *Pascal's triangle*.

Pascal's triangle consists of a sequence of rows, where each row gives the values of the binomial coefficients $\binom{n}{k}$ for a particular value of n. For example, the row with $n = 1$ has only two numbers in it:

$$\binom{1}{0} = 1 \quad \text{and} \quad \binom{1}{1} = 1.$$

The row with $n = 2$ has three:

$$\binom{2}{0} = 1 \quad \binom{2}{1} = 2 \quad \binom{2}{2} = 1,$$

The $n = 3$ row has four:

$$\binom{3}{0} = 1 \quad \binom{3}{1} = 3 \quad \binom{3}{2} = 3 \quad \binom{3}{3} = 1,$$

and so on. These rows are arranged one above the other, with the centers vertically aligned. The whole thing looks like

TABLE 6-1 Pascal's Triangle

								1									$(n = 0)$
							1		1								$(n = 1)$
						1		2		1							$(n = 2)$
					1		3		3		1						$(n = 3)$
				1		4		6		4		1					$(n = 4)$
			1		5		10		10		5		1				$(n = 5)$
		1		6		15		20		15		6		1			$(n = 6)$
	1		7		21		35		35		21		7		1		$(n = 7)$
1		8		28		56		70		56		28		8		1	$(n = 8)$

Of course, the triangle continues forever; but we have to stop somewhere. Note that we start with the row for $n = 0$, consisting of the one binomial coefficient $\binom{0}{0} = 1$. (In general, we'll refer to the row starting with $\binom{n}{0}$, $\binom{n}{1}$, etc., as the nth row.)

As we said, the patterns we've observed so far in the binomial coefficients are all in evidence here. The most striking is the symmetry: if we flip the whole thing around its central vertical axis, the triangle is unchanged. This is, naturally, a reflection of the fact that

$$\binom{n}{k} = \binom{n}{n-k}.$$

We also see that

$$\binom{n}{0} = \binom{n}{n} = 1$$

for any n in the fact that the edges of the triangle consist entirely of 1s, and that

$$\binom{n}{1} = \binom{n}{n-1} = n$$

in the fact that the second (and second-to-last) entry in each row is the row number.

6.2 A New Relation

All the laws we've discussed so far have to do with the entries on a single row: the first and last are 1; the row is symmetric, and so on. But writing out all the binomial coefficients together reveals a pattern among entries in different rows as well. It's obvious once you think of it, and maybe you've seen it before; if you haven't, take a moment out to stare some more at the triangle before we point it out.

OK, here it is: *Each entry in the table is exactly the sum of the two entries closest to it in the row immediately above it.* The entry $\binom{5}{2} = 10$ in the $n = 5$ row is the sum of the two entries $\binom{4}{1} = 4$ and $\binom{4}{2} = 6$ in the $n = 4$ row; the entry $\binom{8}{3} = 56$ in the $n = 8$ row is the sum of the two entries $\binom{7}{2} = 21$ and $\binom{7}{3} = 35$ in the $n = 7$ row, and so on. In fact, you can use this pattern to write down the next ($n = 9$) row of the triangle without actually multiplying and dividing any more factorials, but just adding pairs of terms in the $n = 8$ row:

$$
\begin{array}{ccccccccccccccccccc}
&&&&&&&&&1&&&&&&&&&\\
&&&&&&&&1&&1&&&&&&&&\\
&&&&&&&1&&2&&1&&&&&&&\\
&&&&&&1&&3&&3&&1&&&&&&\\
&&&&&1&&4&&6&&4&&1&&&&&\\
&&&&1&&5&&10&&10&&5&&1&&&&\\
&&&1&&6&&15&&20&&15&&6&&1&&&\\
&&1&&7&&21&&35&&35&&21&&7&&1&&\\
&1&&8&&28&&56&&70&&56&&28&&8&&1&\\
1&&9&&36&&84&&126&&126&&84&&36&&9&&1
\end{array}
$$

Now, all we have so far is a pattern that holds for the first eight rows of the table. That may be pretty convincing, but to a mathematician it's only the first step: having observed this pattern in practice, we now want to express it in mathematical terms and see why (and if) it's always true.

Let's start with the mathematical expression. In the examples we cited a moment ago, we said that the sum of the second and third entries on the $n = 4$ row was equal to the third entry of the $n = 5$ row, and that the sum of the third and fourth entries on the $n = 7$ row was equal to the fourth entry of the $n = 8$ row. One way to express this pattern in general would be to say that the kth entry of the nth row—that is, $\binom{n}{k}$—is equal to the sum of the $(k-1)^{\text{st}}$ and k^{st} entries of the $(n-1)^{\text{st}}$ row. That is,

$$\binom{n}{k} = \binom{n-1}{k-1} + \binom{n-1}{k}.$$

OK, there's our formula; now: is it true? Actually, it is, and not only that but we have two different ways of seeing it! We can either think of the binomial coefficients as solutions of counting problems, and try to understand in that way why this equation might be true; or we can work with the factorial formula for the binomial coefficients—that is, the formula $\binom{n}{k} = \frac{n!}{k!(n-k)!}$—and try to manipulate the sum on the left of this equation to see whether it's equal to the quantity on the right.

Let's start with the interpretation of binomial coefficients as solutions of counting problems. We can best describe this approach first by example. Consider for a moment the number of possible bridge hands—that is, collections of 13 cards chosen without repetition from a deck of 52. We know how many such hands there are; it's just the binomial coefficient

$$\binom{52}{13}.$$

But now suppose we divide all bridge hands into two classes: those that include the ace of spades, and those that don't. How many of each kind are there? Well, hands that don't include the ace of spades are just collections of 13 cards chosen without repetition from among the other 51 cards of the deck, so that there are

$$\binom{51}{13}$$

such hands. Similarly, hands that do include ♠A will consist of the ace of spades plus 12 other cards chosen from among the 51 remaining cards of the deck, so that there are

$$\binom{51}{12}$$

of them. Since every hand either does or doesn't contain the ace of spades, the number of all possible bridge hands must equal the number of those that include ♠A plus the number that don't; in other words,

$$\binom{52}{13} = \binom{51}{12} + \binom{51}{13}.$$

The same logic can be applied for any k and n: if we want to count collections of k objects chosen without repetition from a pool of n, we can single out one particular element of our pool of n (it won't matter which one we pick), and break up all collections of k objects from the pool of n into those that do contain the distinguished element, and those that don't. Those that don't, correspond to collections of k objects chosen from among the remaining $n - 1$; those that do, correspond to collections of $k - 1$ objects from among the remaining $n - 1$. Altogether we see that

$$\binom{n}{k} = \binom{n-1}{k-1} + \binom{n-1}{k}$$

—in other words, *each binomial coefficient is the sum of the two above it in Pascal's triangle*.

There is, as we said, a second way of deriving the formula, by algebra. We know that

$$\binom{n-1}{k-1} = \frac{(n-1)!}{(k-1)!(n-k)!} \quad \text{and} \quad \binom{n-1}{k} = \frac{(n-1)!}{k!(n-k-1)!}.$$

Now, we're taught in elementary school that if we want to add two fractions the first thing to do is to make their denominators equal. We can do that here: we can multiply the top and bottom of the fraction $(n-1)!/(k-1)!(n-k)!$ by k to arrive at

$$\frac{(n-1)!}{(k-1)!(n-k)!} = \frac{k}{k} \cdot \frac{(n-1)!}{(k-1)!(n-k)!} = \frac{k \cdot (n-1)!}{k!(n-k)!}$$

and likewise we can multiply the numerator and denominator of the fraction $(n-1)!/k!(n-k-1)!$ by $n-k$ to see that

$$\frac{(n-1)!}{k!(n-k-1)!} = \frac{n-k}{n-k} \cdot \frac{(n-1)!}{k!(n-k-1)!} = \frac{(n-k) \cdot (n-1)!}{k!(n-k)!}.$$

Now we can add them: we have

$$\frac{(n-1)!}{(k-1)!(n-k)!} + \frac{(n-1)!}{k!(n-k-1)!} = \frac{k \cdot (n-1)!}{k!(n-k)!} + \frac{(n-k) \cdot (n-1)!}{k!(n-k)!}$$

We can combine the k and the $n - k$ to get

$$= \frac{n \cdot (n-1)!}{k!(n-k)!},$$

and merging the n with the $(n - 1)!$ we can rewrite this as

$$= \frac{n!}{k!(n-k)!}$$

which we simply recognize as

$$= \binom{n}{k}.$$

So our formula really does hold! That's the algebraic proof.

6.3 More Relations

Here are two more patterns you might observe if you stare at Pascal's triangle long enough.

Suppose first that you decide, on a whim, to add up the binomial coefficients in each row of the triangle. What do you get? The pattern begins to emerge fairly quickly:

$$\begin{aligned}
1 &= 1 \\
1+1 &= 2 \\
1+2+1 &= 4 \\
1+3+3+1 &= 8 \\
1+4+6+4+1 &= 16 \\
1+5+10+10+5+1 &= 32 \\
1+6+15+20+15+6+1 &= 64
\end{aligned}$$

We see, in other words, that the sum of the binomial coefficients in each row is a power of 2; more precisely, the sum of the numbers on the nth row seems to be 2^n.

Why should this be? Well, this is like the last relation, in that once you see it it's not hard to figure out the reason why. There are many ways to think of this, but for fun (and concreteness) let's do the following:

Imagine that we're at a salad bar, which has (say) seven ingredients: lettuce, tomato, onion, cucumber, broccoli, carrots and the ubiquitous tofu cubes. We ask: *how many different salads can we make*?

Well, this is easy enough to answer using the multiplication principle, if we approach the problem the right way. We can start with a simple choice: should we have lettuce in our salad or not? Next we ask if we want tomatoes or not, and so on; all in all, we have to make seven independent choices, each a yes or no decision. By the multiplication principle, then, we see that there are $2^7 = 128$ possible salads. (Note that we're including the option of just saying no to all the ingredients: the empty salad.)

Suppose on the other hand that we ask: how many different salads can we make that have exactly three ingredients? Again, this is a completely simple application of what we know: there are $\binom{7}{3}$ different ways of choosing three ingredients from among seven. How many salads with two ingredients? $\binom{7}{2}$, of course. With four? $\binom{7}{4}$, and so on.

You can see where this is headed: the total number of salads is the sum of the number of salads with no ingredients, the number of salads with one ingredient, the number of salads with two ingredients, and so on. Since we've already established that the total number is 2^7, we conclude that

$$\binom{7}{0} + \binom{7}{1} + \binom{7}{2} + \binom{7}{3} + \binom{7}{4} + \binom{7}{5} + \binom{7}{6} + \binom{7}{7} = 2^7.$$

What's more, you can see that the same idea will work to establish our relation for any n: just imagine a salad bar with n ingredients, and again ask how many salads we can make. On the one hand, the total number is 2^n by the multiplication principle; on the other, it's equal to the sum of the number $\binom{n}{0}$ of salads with no ingredients, the number $\binom{n}{1}$ of salads with 1 ingredient, the number $\binom{n}{2}$ of salads with 2 ingredients, and so on. Thus the sum of all the binomial coefficients with n on top must equal 2^n.

There's an interesting difference, by the way, between this relation and the one we derived in the preceding section. In that case, we had two different ways of verifying the relation: *combinatorially*, that is, via the interpretation of the binomial coefficients as solutions of counting problems; and *algebraically*, that is, by manipulating the factorial formula for binomial coefficients. In the present case, we do have a fairly straightforward combinatorial way of seeing why the formula should be true. But it's not at all clear from the formulas why it holds: if you didn't know that binomial coefficients count collections—if you had only algebra to work with—it would be hard at this point to show that

$$\frac{n!}{0!n!} + \frac{n!}{1!(n-1)!} + \frac{n!}{2!(n-2)!} + \cdots + \frac{n!}{(n-1)!1!} + \frac{n!}{n!0!} = 2^n.$$

In fact, we will see another way to derive this relation, as well as the following one, from the binomial theorem discussed in the next section.

The next pattern is perhaps more subtle. We're going to look at what's called the *alternating sum* of the binomial coefficients on each row: that is, we're going to look at a particular row and take the first number on that row, minus the second, plus the third, minus the fourth, plus the fifth, and so on to the end. What do we get? Again, the pattern doesn't take long to appear:

$$
\begin{aligned}
1-1 &= 0\\
1-2+1 &= 0\\
1-3+3-1 &= 0\\
1-4+6-4+1 &= 0\\
1-5+10-10+5-1 &= 0\\
1-6+15-20+15-6+1 &= 0
\end{aligned}
$$

As before, we ask if the alternating sum of the numbers in each row is always going to be 0, and if so why. Note that half the time this is obvious: in the $n = 5$ row, for example, each number appears twice, once with a plus sign and once with a minus, so of course they cancel. The same is true in each row with an odd n. But it's much less clear why this should be true in the even rows as well.

Again, there is a counting-based reason why it should be true. We'll try to phrase it in terms of probability. The key question is: if we flip a coin, say, six times, *what is the probability that we'll get an even number of heads*? Likewise, what is the probability that we'll get an odd number of heads?

Well, first off we can answer these questions in terms of the formulas we've derived in the last few chapters. The probability of getting an even number of heads is simply the sum of the probability of getting no heads, the probability of getting two heads, the probability of getting four heads and the probability of getting six heads. We've worked all these out in the last chapter; the answer is

$$\frac{\binom{6}{0}+\binom{6}{2}+\binom{6}{4}+\binom{6}{6}}{2^6},$$

and similarly the probability of getting an odd number of heads is

$$\frac{\binom{6}{1}+\binom{6}{3}+\binom{6}{5}}{2^6}.$$

So far, so good. But there's another way to think about the same problem that gets us directly to the answer. It's simply the observation that *it all comes down to the last flip*: whatever the outcome of the first 5 flips, the question of whether the total number is heads will be odd or even depends on the outcome of the last flip. That is, if the number of heads on the first five flips is even, the total for all six will be even if the last flip is a tail and odd if it's a head; if the number after five flips is odd, the total will be even if the last flip is a head and odd if it's a tail. Either way, the odds are 50-50: the probability of the total number being odd is equal to the probability that the total will be even. But we've already worked out what those odds are, and the conclusion is that the two expressions above must be equal; that is,

$$\binom{6}{0}+\binom{6}{2}+\binom{6}{4}+\binom{6}{6}=\binom{6}{1}+\binom{6}{3}+\binom{6}{5}$$

But this is exactly to say that the alternating sum

$$\binom{6}{0}-\binom{6}{1}+\binom{6}{2}-\binom{6}{3}+\binom{6}{4}-\binom{6}{5}+\binom{6}{6}=0,$$

and so we've established our relation in this case. Moreover, it's pretty clear the same logic will work for any n: just imagine we're flipping a coin n times and again ask what the probability is of getting an even number of heads.

We're going to stop here, even though there are many many more patterns to discern in Pascal's triangle. (We'll mention one more in Exercise 6.3.2 below.) It's worth pointing out, though, that these arguments both illustrate a principle that's true in many aspects of life outside mathematics: the key to finding a good answer is to ask the right question.

Here's another way to think about the two relations we've just seen:

Exercise 6.3.1 Use the fact that each number in Pascal's triangle is the sum of the two immediately above it to convince yourself that the alternating sum of the numbers in each row is 0. Can you make a similar argument for the fact that the sum of the numbers in the nth row is 2^n?

Here's one more:

Exercise 6.3.2 Try this: starting with any of the 1s on an edge of Pascal's triangle, and moving along a line parallel to the opposite edge, add up all the numbers until you get to a particular row. For example, if you start with the binomial coefficient $\binom{2}{2} = 1$ and continue to the $n = 6$ row we'd get the sum

$$\binom{2}{2} + \binom{3}{2} + \binom{4}{2} + \binom{5}{2} + \binom{6}{2};$$

or, more graphically, the sum of the boxed numbers in the triangle below

$$\begin{array}{ccccccccccccccccc}
&&&&&&&&1&&&&&&&& \\
&&&&&&&1&&1&&&&&&& \\
&&&&&&1&&2&&\boxed{1}&&&&&& \\
&&&&&1&&3&&\boxed{3}&&1&&&&& \\
&&&&1&&4&&\boxed{6}&&4&&1&&&& \\
&&&1&&5&&\boxed{10}&&10&&5&&1&&& \\
&&1&&6&&\boxed{15}&&20&&15&&6&&1&& \\
&1&&7&&21&&35&&35&&21&&7&&1& \\
1&&8&&28&&56&&70&&56&&28&&8&&1
\end{array}$$

What do these numbers add up to in this case? In general? Try this a few times and see if you spot the pattern; then see if you can convince yourself that this pattern always holds.

6.4 The Binomial Theorem

The binomial theorem is concerned with powers of the sum of two numbers. With some trepidation—as you've seen, in this book we try to work with numbers rather than letters when we can—we'll call them x and y. The question is, simply, what do we get when we take the sum $x + y$ and raise it to a power?

Let's start with the first example: let's multiply $x + y$ by itself. (Of course we know you already know how to do this, but bear with us: we want you to think about what you're doing when you do it.) Write it out as:

$$\begin{array}{rr} & x + y \\ \times & x + y \\ \hline \end{array}$$

When we multiply this out, we're going to get four terms: we have x times itself, or x^2; we have y times itself, and we have two cross terms: an x times a y and a y times an x. Altogether, it adds up to $x^2 + 2xy + y^2$.

Next, consider what happens when we raise $x + y$ to the third power. Again, we'll write it out as a product:

$$\begin{array}{rr} & x + y \\ & x + y \\ \times & x + y \\ \hline \end{array}$$

Try to anticipate what you're going to get when you multiply this out. First of all, count the number of terms you're going to get: when you expand, you have to pick one of the two terms from each factor $x + y$ and multiply them out; by the multiplication principle there'll be 2^3, or eight, terms. Now, you're going to get one term x^3, the product of the three x's. The next question is: how many terms are you going to get that are a product of two x's and a y? The answer is three: you can choose the x term from any two of the three factors, and the y term from the other. Likewise, you're going to get three terms in the product involving two y's and an x, and one y^3 term; altogether we see that

$$(x + y)^3 = x^3 + 3x^2y + 3xy^2 + y^3.$$

You can probably see where this is headed already, but let's do one more example. Consider the fourth power $(x + y)^4$:

$$\begin{array}{rr} & x + y \\ & x + y \\ & x + y \\ \times & x + y \\ \hline \end{array}$$

When we multiply out this product, as before we're going to get one term that is the product of the four x's. Likewise, there are going to be four terms involving three x's and a y: we can pick the x term from three of the factors and the y term from the fourth. And similarly, the number of terms in the expanded product that involve two x's and two y's will be the number of ways of picking two of the four factors in the product: that is, $\binom{4}{2}$, or six. All in all, we see that

$$(x + y)^4 = x^4 + 4x^3y + 6x^2y^2 + 4xy^3 + y^4.$$

The picture in general is just what you'd expect. When we multiply out the nth power of $x + y$, we get a total of 2^n terms, exactly $\binom{n}{k}$ of which will be the product of k x's and $(n - k)$ y's. What this means is that we have a general formula

$$(x + y)^n = \binom{n}{0}x^n + \binom{n}{1}x^{n-1}y + \binom{n}{2}x^{n-2}y^2 + \cdots + \binom{n}{n-1}xy^{n-1} + \binom{n}{n}y^n;$$

or, in English:

> The coefficient of $x^k y^{n-k}$ in $(x+y)^n$ is $\binom{n}{k}$.

This result is called the Binomial Theorem, and it's where the binomial coefficients got their name.

One last item: we promised to show you two new ways to see the relations we worked out in the last section. In fact, they're easy to do once we have the binomial theorem.

The point is, the formula for $(x+y)^n$ we just wrote down is an algebraic equation: that is, it's valid if we substitute any two numbers for x and y. So, for example, what happens if we substitute 1 for x and 1 for y? Well, $x + y = 2$, as we don't need to tell you; and $x^k y^{n-k} = 1$ no matter what n and k are (likewise, we hope). So the general formula above now reads

$$2^n = \binom{n}{0} + \binom{n}{1} + \binom{n}{2} + \cdots + \binom{n}{n-1} + \binom{n}{n}.$$

which was our first relation.

Similarly, we could plug in 1 for x and -1 for y. This time the sum $x + y$ is zero; and the products $x^k y^{n-k}$ are alternately 1 and -1: $x^n = 1$, $x^{n-1}y = -1$, $x^{n-2}y^2 = 1$ and so on. Now when we substitute we arrive at

$$0 = \binom{n}{0} - \binom{n}{1} + \binom{n}{2} - \binom{n}{3} + \cdots .$$

That's our second relation.

Before doing the next two exercises, you might want to review the discussion of multinomial coefficients in Section 4.4.

Exercise 6.4.1 What is the coefficient of $x^3y^2z^3$ in $(x+y+z)^8$? (If you understood the derivation of the binomial theorem, this should be easy.)

Exercise 6.4.2 What is the sum of all multinomials

$$\binom{7}{a, b, c}$$

with a 7 on top, and any three numbers adding up to 7 below?

Chapter 7

Advanced Counting

As we've said, if you've gone through the first five chapters of this book, you have a pretty good foundation in counting—more than enough, in particular, to see you through the rest of the book. Nonetheless, in the final chapter of this part we'd like to take up two additional topics, just to give you an idea of some of what's out there.

Actually, the first of the two topics isn't that much more advanced than what we've already done—it's the formula for the number of collections with repetitions, as discussed in Section 4.5. In fact, the problem is pretty much on a par with the ones we've looked at up to now; it's the derivation, rather than the formula itself, that might be considered less elementary than the content of this book so far.

The second topic—Catalan numbers and some of their applications—is just a really cute exercise in more advanced counting. If you've enjoyed the challenges so far in this part of the book, you may get a kick out of seeing how to tackle a new sort of problem; if not, you could approach it with an anthropologist's point of view: it'll at least give you an idea of some of the more esoteric things mathematicians love to count.

7.1 Back to the Fruit Bowl

Let's start by recalling briefly the discussion in Section 4.5. In the body of this part of the book, we've derived three main formulas.

- We've counted the number of ways of choosing a sequence of k objects (that is, a succession of choices where the order of selection matters) from a common pool of n objects, with repetition allowed: for example, words of a given length, outcomes of k coin flips or dice rolls, and so on.
- We've counted the number of ways of choosing a sequence of k objects from a common pool without repetition: for example, words of a given length with no repeated letters, class officers, etc. And
- We've calculated the number of ways of choosing a collection of objects (that is, a succession of choices where the order of selection doesn't matter) from a pool of n objects, without repetition: committees chosen from the pool of students in a class, video rentals and the like.

If we arrange these in a table, as we did in Section 4.5

	repetitions allowed	without repetitions
sequences	n^k	$\frac{n!}{(n-k)!}$
collections	??	$\frac{n!}{k!(n-k)!}$

we see that there's an obvious gap: we don't know how to count the number of ways of choosing a collection of objects from a common pool, *with repetition allowed*. It's time to remedy the situation.

Let's start with an example: the fruit bowl introduced in Section 4.5. (We'll change the numbers to make the problem a little more manageable.) The situation is this: you have a fruit bowl, containing (say) eight varieties of fruit—apples, bananas, canteloupes, durian, elderberries, figs, grapefruit and honeydews—with unlimited quantities of each. You're assembling a little snack; you figure five servings of fruit would be about right. The question is, how many different snacks can you choose?

Now, if for some reason you decided to rule out taking more than one serving of any one fruit—in other words, if you wanted to consider only snacks consisting of five *different* fruits—this would be a standard problem: we'd be looking at collections of five objects from among eight, without repetition, and the answer would be $\binom{8}{5}$, or 56. But we're not ruling out taking more than one piece of a given fruit; you might decide to go with three apples and two bananas, for example, or this could be a five-banana day. So there should be more possibilities. How many?

Before we go ahead and solve this problem, let's consider some approaches that don't work. For example, we might try the trick we used to count collections without repetition, in Section 4.2: we might first try to count sequences of fruits, rather than collections. (It's a stretch, but imagine we were making a menu for our snack, in which we specified not only the fruits but the order in which they were to be consumed; we could ask how many possible menus there were.) As in Section 4.2, the number of sequences, or menus, is a standard calculation: there are 8^5, or 32,768.

But here the method breaks down. In the case of sequences without repetition, each collection of five objects corresponded to exactly 5!, or 120, different sequences, corresponding to the different orders in which you could eat the five different fruits. But when we allow repetitions, not all collections give rise to 120 different menus: if you went with the 3 apples and 2 bananas, for example, the number of possible menus—the number of ways of ordering the five—would be $\binom{5}{3}$, or 10, corresponding to the number of ways of positioning the apples on your menu. And of course if you decided on the five bananas there would be only one possible menu. So we can't just divide the number of sequences by 120 to arrive at the number of collections, as we did in the no-repetition case.

OK, how about this? We know that there are $\binom{8}{5} = 56$ collections of five fruits without repetition. How many collections are there with two of one fruit and one each of three others? We can do this as well: pick the one fruit we want to double up on—that's eight choices—then choose three fruits without repetition from among the seven fruits remaining—that's $\binom{7}{3} = 35$ choices, for a total of

$$8 \times 35 = 280$$

possibilities. Likewise, we can count the number of collections with three of one fruit and one each of two others (that's $8 \times \binom{7}{2}$, or 168); with two each of two fruits and one of another ($\binom{8}{2} \times 6 = 168$), and so on. We can, in fact, consider in turn all the possible ways of assembling 5 fruits and add up the number of ways of doing each. But this is a very cumbersome method, and doesn't lead to a simple formula. And, as you might expect, when the numbers get a little larger it becomes completely intractable.

The solution, finally, requires us to approach the problem a little differently: we have to think *graphically* rather than numerically. This is an approach that we haven't seen up to now, but it's something mathematicians do a lot, especially when dealing with counting problems. The general idea is to associate to each of the objects we're trying to count a graphical object—a diagram, a collection of blocks or whatever—and then count the number of such diagrams that arise. Probably the best way to explain this approach is to illustrate it with examples, and we'll see examples of it both in this section and in the derivation of the formula for the Catalan numbers in Section 7.6.

This sort of approach often requires some ingenuity: it's usually not clear in advance what sort of graphical objects to associate to the things we want to count. In the present circumstance—counting collections of fruits—what we're going to do is to represent each possible choice by a *block diagram*, as we'll describe in a moment. It may not be obvious at first why we're doing this, but (at the risk of mixing food groups) the proof of the pudding is in the eating: at the end, we'll see that we've converted the problem into one we can solve readily.

On to the problem! To begin with, suppose that the fruits are alphabetized: we have apples, bananas, canteloupes and so on up to grapefruit and honeydew melons. Now say that we had five white boxes, corresponding to the possible choices of fruits, and seven blue, or "divider," boxes, seven being here one less than the number of different fruits we're choosing among. Next, suppose we have in mind a particular choice of snack. We can represent that choice by arranging the boxes in a row according to the following rule:

- The number of blank boxes to the left of the first divider is the number of apples,
- The number of blank boxes between the first and second dividers is the number of bananas,
- The number of blank boxes between the second and third dividers is the number of canteloupes,

  and so on, until
- The number of blank boxes between the sixth and seventh dividers is the number of grapefruit, and finally
- The number of blank boxes to the right of the last divider is the number of honeydew melons.

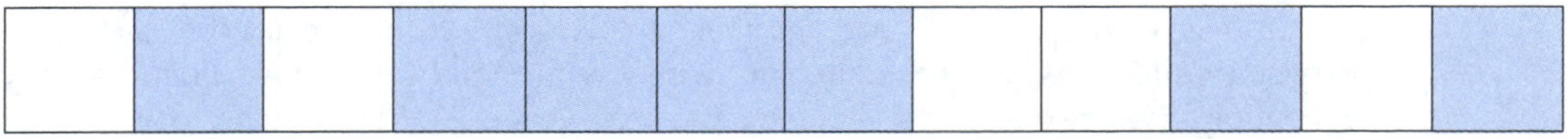

Thus, for example, the choice of "an apple, a banana, two figs and a grapefruit" would correspond to the diagram while the selection "one canteloupe, one durian, one fig and two honeydew melons" would be represented by the diagram

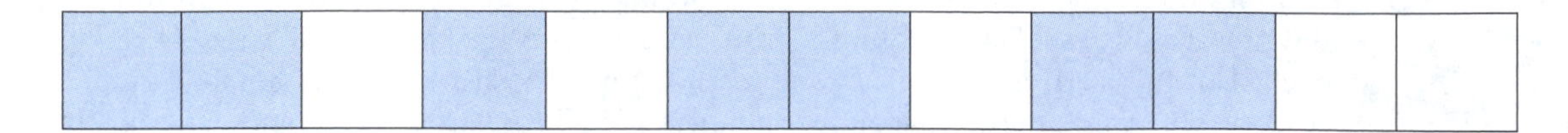

Thus, every selection may be represented by a diagram of 12 boxes, seven of them blue and five white. Conversely, given any such diagram, we can read off from it a choice of fruit: for example, to the diagram

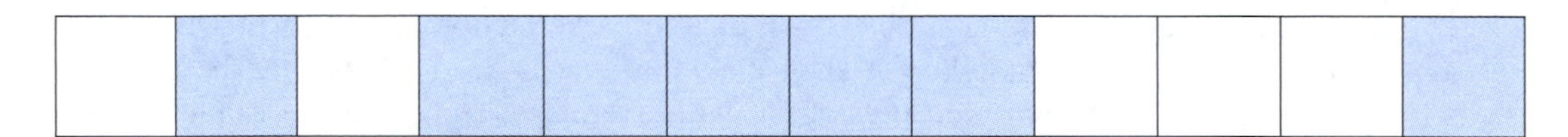

corresponds to the choice, "one apple, one banana and three grapefruit."

This construction may have seemed somewhat arbitrary when we first suggested it, but by now the reason should be becoming clearer: what we have is an exact correspondence between ways of choosing a collection of five objects from among eight, with repetition, and box diagrams. The number of collections is thus the same as the number of box diagrams, and this is something we know how to count: it's simply the number of ways of choosing seven boxes from among 12 to color in, and that is

$$\binom{12}{7} = 792.$$

In other words, the number of choices of fruit snacks is the number of possible positions of the seven dividers we need to put in the 12 slots to separate them.

In a similar fashion we can solve the problem of counting collections of k objects from among n in general. We begin by ordering the objects in our pool arbitrarily: we'll call them "object #1," "object #2" and so on up to "object #n." Then, to any such collection, we associate a box diagram, with $n - 1$ blue, or divider boxes, and k white boxes. The rule, as in the fruit example, is simple:

- The number of blank boxes to the left of the first divider is the number of times object #1 is chosen;
- The number of blank boxes between the first and second dividers is the number of times object #2 is chosen;

  and so on until
- The number of blank boxes between the $(n - 2)^{\text{nd}}$ and $(n - 1)^{\text{st}}$ dividers (that is, the last two) is the number of times object #$(n - 1)$ is chosen; and finally
- The number of blank boxes to the right of the last, or $(n - 1)^{\text{st}}$, divider is the number of times object #n is chosen.

Just as in our example, we see that in this way we set up a correspondence between collections and box diagrams with k white and $n - 1$ blue boxes—every collection gives rise to a box diagram, and vice versa. And the number of such box diagrams is one we know; it's just the number of ways of saying where the k white boxes should go among the $n+k-1$ slots in the diagram. The conclusion, in other words, is that

> The number of ways of choosing a collection of k objects from among n objects, with repetitions allowed, is
>
> $$\binom{n+k-1}{k} = \frac{(n+k-1)!}{k!(n-1)!}.$$

We now know how to count both sequences and collections of objects chosen from a common pool, both with and without repetitions. In particular, we can complete the diagram we introduced in Section 4.5:

	repetitions allowed	*without repetitions*
sequences	n^k	$\frac{n!}{(n-k)!}$
collections	$\binom{n+k-1}{k}$	$\binom{n}{k}$

Exercise 7.1.1 Just to do this once: in the fruit example at the beginning of this section, calculate the number of choices consisting of

1. five different fruits
2. two of one fruit and one each of three others
3. two each of two fruits and one of another
4. three of one fruit and one each of two others
5. three of one fruit and two of another
6. four of one fruit and one of another
7. five of one fruit.

Add these up. Does the result agree with the answer obtained above? (If not, do the problem again.)

Exercise 7.1.2 Suppose again that you're the chief distributor for the National Widget Manufacturing Co. The NWMC has one central widget-producing plant, and five distribution centers.

Let's say the central plant has just produced 12 crates of widgets, and it's your job to say how many of these 12 each distribution center is to receive. How many ways are there of doing this?

Exercise 7.1.3 By a *rack* in the game of Scrabble we'll mean a collection of seven letters; the order doesn't matter, and of course letters can be repeated. How many different racks are possible? (It's not the case in practice, but for the purposes of this calculation, let's say there are at least seven of each letter available.)

Exercise 7.1.4 A company wants to place eight orders for widgets with 13 suppliers.

1. How many ways are there of placing the orders if no supplier is to be given more than one order?
2. How many ways are there of placing the orders if any supplier can be given any number of orders?
3. Of the suppliers, seven are in-state and six are out of state. How many ways are there of placing the orders if no supplier is to be given more than one order and at least 6 orders must be placed in-state?

7.2 Catalan Numbers

Let's start with a strange-sounding question: in how many ways can a collection of parentheses appear in a sentence? Say, for example, that a sentence has exactly one pair of parentheses. Ignoring the rest of the text, there's no question: the parentheses must appear in the order (). (The one restriction on parentheses is that you can't close a parenthetical statement that hasn't yet been opened.)

But suppose now that a sentence has two pairs of parentheses. It might have two separate parenthetical statements, so that the parentheses appear like this:

$$()()$$

or it could have one parenthetical statement included in another, so that if you stripped away all the text the parentheses would be in this order:

$$(())$$

These are the only possibilities, so we would say there are two ways for two pairs of parentheses to appear in a sentence.

What about three pairs? At this point, you should stop reading and try to work out on your own how many ways three pairs of parentheses might appear in a sentence; we'll tell you the answer on the next page.

STOP.

CLOSE THE BOOK.

GRAB A PAD OF PAPER AND A PEN.

WORK OUT SOME EXAMPLES ON YOUR OWN.

THINK.

The answer is that there are *five* ways they might appear:

()()()   ()(())   (())()   (()())   and   ((()))

Of course, we're talking mathematics here, not literary style. You won't find too many sentences (in nontechnical writing at least) with three pairs of parentheses arranged as in the second of these configurations (and if you did it would (probably) be a fairly convoluted sentence).

For four pairs it turns out there are 14 ways:

(((())))   ((()()))   ((())())   ((()))()   (()(()))   (()()())   (())(())

(())(())   (())()()   ()((()))   ()(()())   ()()(())   ()(())()   and   ()()()()

In general, we ask: in how many ways can n pairs of parentheses appear in a sentence? This number is called the nth *Catalan number*, and is denoted c_n. By convention, we say that the 0th Catalan number c_0 is 1. Thus we have seen that

$$\begin{aligned} c_0 &= 1 \\ c_1 &= 1 \\ c_2 &= 2 \\ c_3 &= 5 \\ c_4 &= 14. \end{aligned}$$

The next few Catalan numbers are

$$\begin{aligned} c_5 &= 42 \\ c_6 &= 132 \\ c_7 &= 429. \end{aligned}$$

The Catalan numbers are a fascinating sequence of numbers that arise in a surprising variety of counting problems. (They were named after Eugène Catalan, a 19th century Belgian mathematician, who encountered them in counting the number of ways you could dissect an n-sided polygon into triangles.) In the remainder of this chapter, we'll mention some of these ways, but the main thing we want to do is to describe a pair of formulas for the Catalan numbers. The first is what we call a *recursive formula*: it expresses each Catalan number in terms of all the previous ones. The second (which is somewhat harder to derive) is by contrast a *closed formula*, which is to say it gives a way of calculating any Catalan number c_n directly; you don't need to know the preceding Catalan numbers to apply it.

7.3 A Recursion Relation

The Catalan numbers, as we said, satisfy a *recursion relation*: that is, there is a rule for finding each Catalan number if you know all the preceding ones. This relation is relatively straightforward to see, starting from the description of the Catalan numbers in terms of parentheses.

To derive it, let's go back to the list of the 14 ways that four pairs of parentheses can appear. This time, though, let's try to see if we can list the ways systematically. Of course, any such sequence has to start with a left parenthesis. The key question is: *where in the sequence is its mate?*—that is, the right parenthesis that closes the parenthetical clause it begins. For example, one possibility is that its mate comes right after it: in other words, that the second symbol in the sequence is a right parenthesis, immediately ending the clause begun by the first. What follows then would be simply an arrangement of the remaining three pairs of parentheses, so that the whole sequence would appear as

$$\Big(\Big) \quad \{3 \text{ pairs}\}.$$

The number of such sequences is just the number of arrangements of three pairs of parentheses, that is, $c_3 = 5$. Explicitly, these are:

$$\Big(\Big)()()() \qquad \Big(\Big)()(()) \qquad \Big(\Big)(())() \qquad \Big(\Big)(()()) \qquad \text{and} \qquad \Big(\Big)((())).$$

Next, it could be that between the initial left parenthesis and its mate there is exactly one pair of parentheses, and after its mate there are two pairs: schematically,

$$\Big(\ \{1 \text{ pair}\} \ \Big) \quad \{2 \text{ pairs}\}.$$

Now, there is no choice about the arrangement of the single pair between the initial open parenthesis and its mate; but we do have $c_2 = 2$ choices for the last two pairs. There are thus altogether 2 sequences of this type; explicitly,

$$\Big(()\Big)()() \qquad \text{and} \qquad \Big(()\Big)(()).$$

The third possibility is that between the initial left parenthesis and its mate there are exactly two pairs of parentheses, and after its mate there is one pair: schematically,

$$\Big(\ \{2 \text{ pairs}\} \ \Big) \quad \{1 \text{ pair}\}.$$

As before, there is no choice about the arrangement of the single pair following the initial open parenthesis and its mate; but we do have $c_2 = 2$ choices for the two pairs nested inside them. There are thus altogether 2 sequences of this type; explicitly,

$$\Big(()()\Big)() \qquad \text{and} \qquad \Big((())\Big)().$$

Finally, the last possibility is that the mate of the initial open parentheses is the final right parenthesis: in other words, the entire sequence is part of one parenthetical clause, or schematically,

$$\Big(\ \{3 \text{ pairs}\} \ \Big).$$

There are $c_3 = 5$ such sequences, corresponding to the five possible arrangements of the three pairs of parentheses in the middle; explicitly,

$$\Big(()()()\Big) \qquad \Big(()(())\Big) \qquad \Big((())()\Big) \qquad \Big((()())\Big) \qquad \text{and} \qquad \Big(((()))\Big).$$

We add up all these possibilities, and we see that $c_4 = 5 + 2 + 2 + 5 = 14$.

We can use the same approach to calculate any Catalan number c_n assuming we know all the ones before it. Basically, we do exactly what we've just done here for c_4: we break up all possible sequences according to where the mate of the initial parenthesis appears—that is, how many pairs appear between them. In other words, every sequence of n pairs of parentheses has schematically the form

$$\Big(\ \{i \text{ pairs}\} \ \Big) \quad \{n - i - 1 \text{ pairs}\}$$

for some i (including $i = 0$ and $i = n - 1$ as possibilities). To specify a sequence of this form we have to choose one of the c_i ways of arranging the i pairs nested inside the initial parenthesis and its mate, and to choose one of the c_{n-i-1} arrangements of the parentheses following. There are thus $c_i \cdot c_{n-i-1}$ sequences of this form, and so we arrive at the recursive formula

$$c_n = c_0 c_{n-1} + c_1 c_{n-2} + c_2 c_{n-3} + \cdots + c_{n-2} c_1 + c_{n-1} c_0.$$

To express this in words: suppose you write out the Catalan numbers from c_0 to c_{n-1}—for example, if $n = 5$, this would be the sequence

$$1 \quad 1 \quad 2 \quad 5 \quad 14$$

Now write the same sequence in reverse directly below this one:

$$\begin{array}{ccccc} 1 & 1 & 2 & 5 & 14 \\ 14 & 5 & 2 & 1 & 1 \end{array}$$

Now multiply each pair of numbers lying directly over one another:

$$14 \quad 5 \quad 4 \quad 5 \quad 14$$

and add up these products:

$$14 + 5 + 4 + 5 + 14 = 42.$$

The result will then be the next Catalan number—in this case, $c_5 = 42$. Which you have to admit is a lot simpler than writing out all 42 ways.

Exercise 7.3.1 Check that this rule works for the first four Catalan numbers (starting with c_1), and use it to calculate the next Catalan number c_6.

7.4 Another Interpretation

Here is another interesting interpretation of the Catalan numbers. Remember when we first introduced binomial coefficients, we described the binomial coefficient $\binom{k+l}{k}$ as the number of possible paths leading from the lower left to the upper right corner of a $k \times l$ grid, where at each junction the path went either up (north) or to the right (east). The reason was, we could describe such a path by a sequence of letters consisting of k Ns and l Es: a sequence such as

E N E E N E N

would correspond to the directions, "go right, then up, then right and right again, then up, then right and finally up," or in other words to this path:

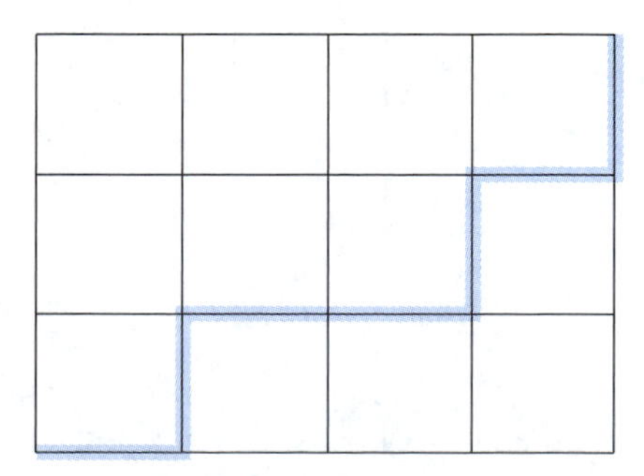

Now imagine that we have n pairs of parentheses appearing in a sentence—say

$$(()()()).$$

Suppose we replace each open parenthesis with an N and each close parenthesis with an E. The resulting sequence

N N E N E N E E

corresponds to a set of directions taking you from one corner to the opposite corner of a 4×4 grid:

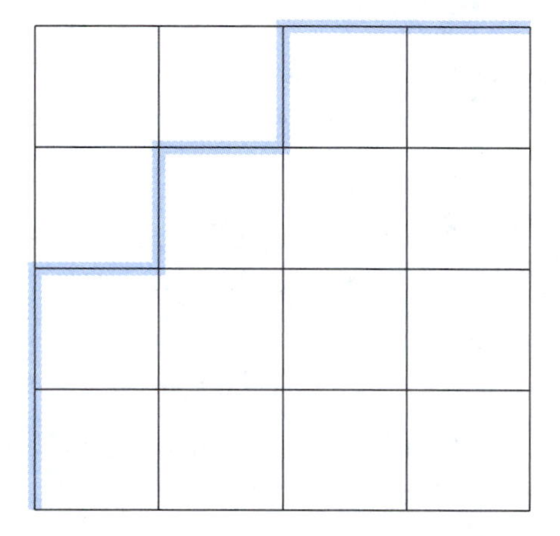

But what does it mean to say that these directions correspond to a "grammatical" sequence of parentheses? If you think about it, it means you can't close a pair of parentheses before you open them—that is, at every point in the sequence you must have had at least as many open (left) parentheses as close (right) parentheses. In terms of the path corresponding to the sequence, this just means that at every stage you must have gone at least as far north as east. In other words, *the path must stay above the diagonal*, though it may touch it. We can thus say that *the* n^{th} *Catalan number* c_n *counts the number of paths from one corner of an* $n \times n$ *triangle to the other*. For example, the number of paths through the grid

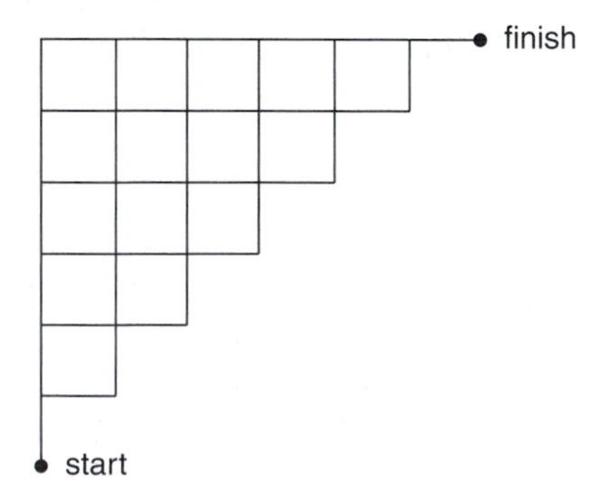

would be the 6th Catalan number c_6.

Exercise 7.4.1 Draw the $c_3 = 5$ possible paths that stay above the diagonal on a 3×3 grid, and the $c_4 = 14$ possible paths that stay above the diagonal on a 4×4 grid.

7.5 The Closed Formula

Consider for a moment *all* possible sequences of n left parentheses and n right parentheses, without regard to logic or grammar. We know that the number of such sequences is just the number of ways of placing the n left parentheses among the $2n$ parentheses altogether; that is, it's equal to the binomial coefficient $\binom{2n}{n}$. We might phrase the question: among all ways in which n left parentheses and n right parentheses might appear in a sequence, *what fraction of these ways are grammatically possible*? This suggests an experiment: let's compare the Catalan numbers 1, 1, 2, 5, 14, 42, ... with the corresponding binomial coefficients, and see if we can see a pattern in their ratios. Here are the results:

n	c_n	$\binom{2n}{n}$	ratio
0	1	1	1:1
1	1	2	1:2
2	2	6	1:3
3	5	20	1:4
4	14	70	1:5
5	42	252	1:6

Well, that's clear enough: based on the evidence so far, it seems natural to think that the nth Catalan number c_n is given by the formula

$$c_n = \frac{1}{n+1}\binom{2n}{n}.$$

Is it? (Check it for the sixth Catalan number you found in Exercise 7.3.1.) In fact, this formula holds in general, and we'll see why in just a moment. Before we get there, though, you should think about it first on your own: why should this formula hold? It's a fun problem to think about.

One way to express this formula is to say that, of the $\binom{2n}{n}$ paths leading from the lower left to the upper right corner of an $n \times n$ grid, the fraction of those that stay above the diagonal is $\frac{1}{n+1}$. If you think in those terms, you'll be able to solve the following problem:

Exercise 7.5.1 Let's say 20 people show up to go see the \$5 matinee at the local theater. Suppose that 10 of these people have exact change, but the other 10 have only a \$10 bill, and will require change. Unfortunately, on this particular day the cashier has forgotten to stop at the bank and so he has no change to start with. What are the odds that he will be able to sell all 20 people tickets without running out of change?

7.6 The Derivation

We're going to finish this chapter—and this part of the book—by showing you how to count the Catalan numbers from scratch. Like the derivation of the formula for collections with repetitions that we showed you in the first section of this chapter, this will be trickier—less straightforward—than some of the other things we've done so far. Like that derivation, also, it will rely in a fundamental way on a graphical representation of the objects being counted.

Let's start with the interpretation of both Catalan numbers and binomial coefficients as paths through a grid. Suppose, for example, we want to look at paths going five blocks north and five blocks east on a rectangular grid—that is, between the two marked points in the figure below. (We've actually enlarged the displayed grid here, for reasons that will become apparent in moment, but we're still interested just in paths going between the two marked points.)

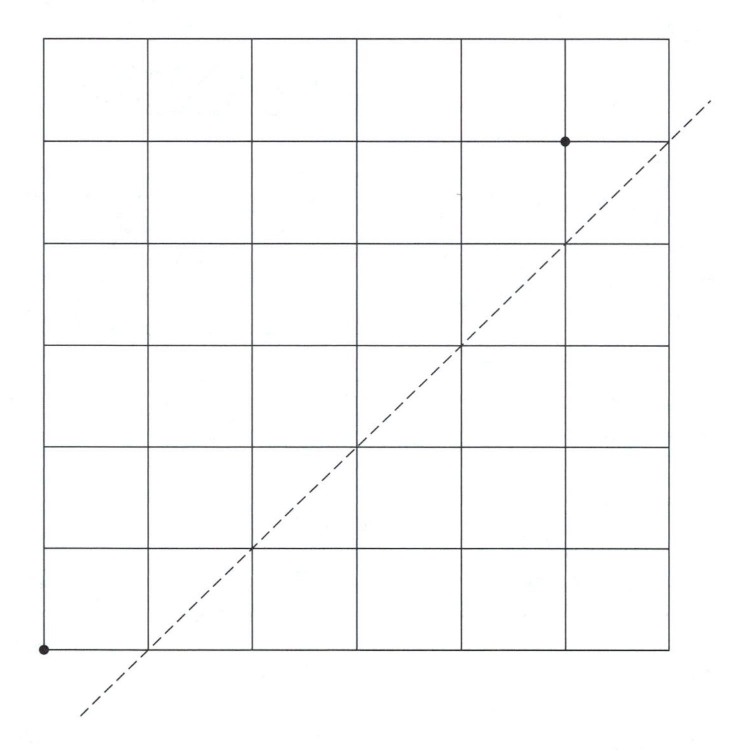

There are $\binom{10}{5} = 252$ such paths altogether; the Catalan number $c_5 = 42$ represents, as we've said, the number of such paths that *do not touch or cross the dotted line*. How do we count such paths? To start with, the first step is to apply the subtraction principle: the number of paths that don't touch the dotted line will be $\binom{10}{5}$ minus the number that do.

Well, that doesn't seem to get us very far: counting the number of paths that do touch the dotted line seems no easier than counting those that don't. Ah, but it is!—at least, if we're clever enough. Here's the trick: if we have any path that *does* touch the dotted line, consider the first point at which it touches. We then break up the path into two parts: the first part from the starting point up to that first touch, and the second from that touch to the end, and we *replace the second part of the path with its reflection in the dotted line*. To visualize this, imagine flipping the whole plane over by rotating it 180° around the dotted line, and carrying the second half of the path—that is, the part after the first touch—with it. Thus, for example, if we start with the path

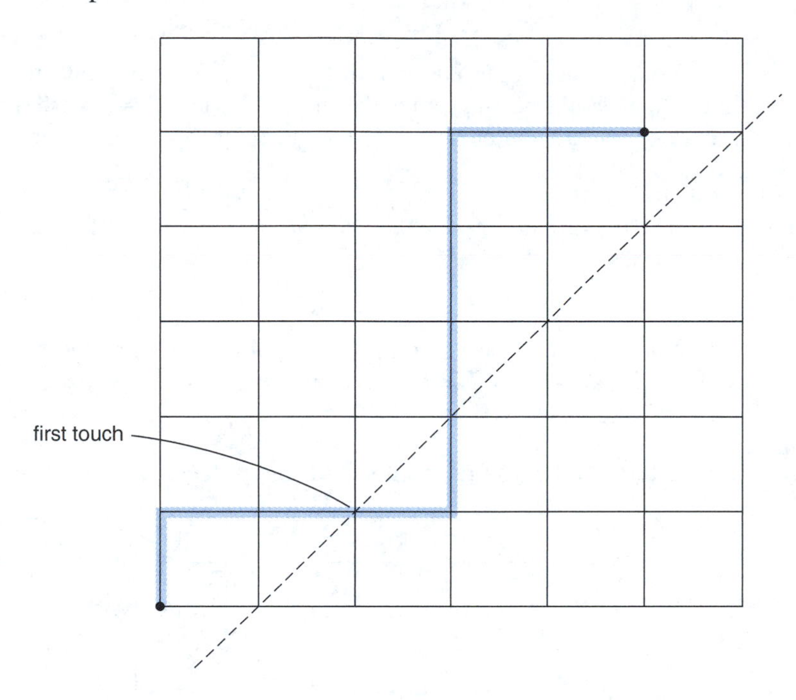

and take the portion of the path after the first touch and reflect in the dotted line, we get

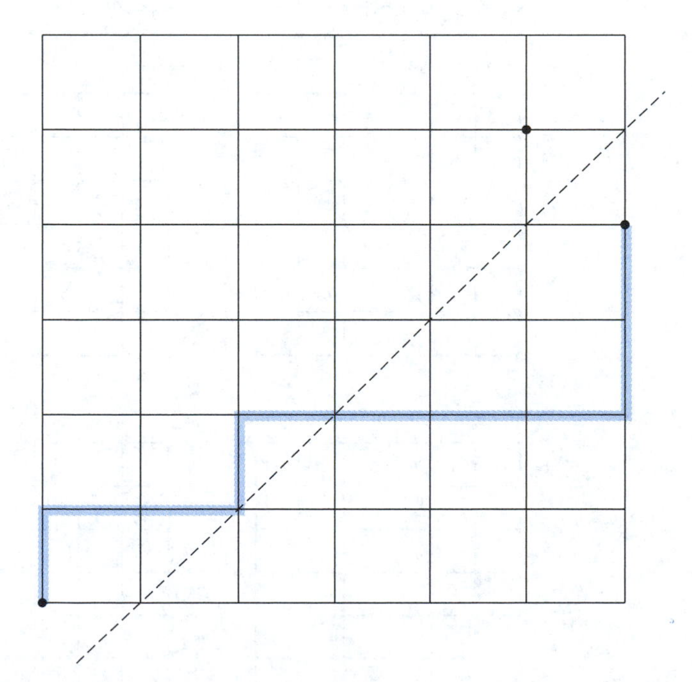

To say it another way: in this example, the original path is represented by the directions

N, E, E, E, N, N, N, N, E, E.

What we are doing is going to the point where the path first touches the dotted line—that would be right after the initial "N, E, E"—and from that point on in the sequence we are switching Ns to Es and Es to Ns to arrive at the sequence

N, E, E, N, E, E, E, E, N, N.

In terms of parentheses, the path we started with corresponds to the sequence of parentheses

()))((((()).

What we are doing is going to the point right after the first illegal parenthesis, and from that point on we are reversing each parenthesis; we arrive at

())()))((.

Here are a few more examples of this process:

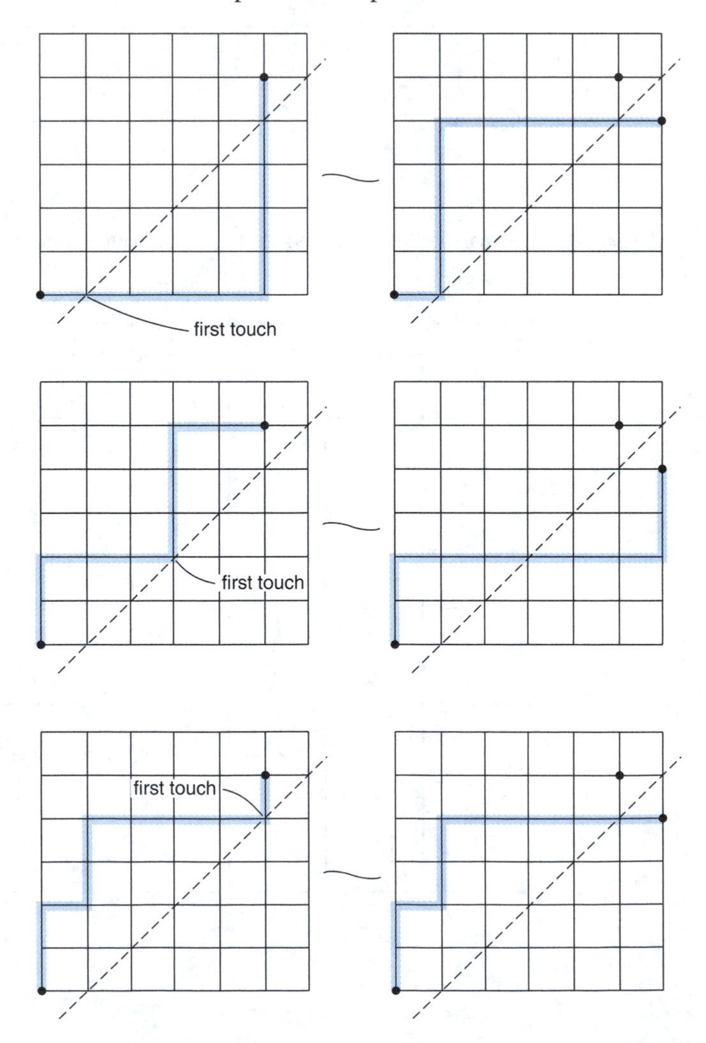

Now, notice one thing about this process. Since the original path goes to the point five blocks north and five blocks east of the starting point, the rejiggered path will go to its mirror image in the dotted line—that is, the point four blocks north and six blocks east of the starting point. Thus, for every path running five blocks north and five blocks east that does touch the dotted line, we arrive at a path going four blocks north and six blocks east.

Moreover, we can always reverse the process. If we have any path going four blocks north and six blocks east, it must cross the dotted line since its destination is on the other side of the dotted line from its origin. We can again look at the point where the path first touches or crosses the dotted line, break the path at that point, and flip the second half of the path around the dotted line. We thus have a correspondence between the two types of paths, and we may conclude that *the number of paths running five blocks north and five blocks east that do touch the dotted line is equal to the number of all paths running four blocks north and six blocks east.*

And that does it: we know the number of all paths running four blocks north and six blocks east is simply the binomial coefficient $\binom{10}{4} = 210$. Thus, of all the $\binom{10}{5} = 252$ paths going five blocks north and five blocks east, 210 do touch the dotted line; it follows that $252 - 210 = 42$ of them don't touch the dotted line, and thus we've calculated the fifth Catalan number $c_5 = 42$.

What's more, exactly the same analysis may be applied to calculate all the Catalan numbers! For any number n, we can draw a similar grid, again with a dotted line one block east of the diagonal. Again, we want to count the paths running n blocks north and n blocks east that stay on or above the diagonal; we do it by counting those that dip below the diagonal (that is, that touch the dotted line) and subtracting this from the total number of paths running n blocks north and n blocks east. And finally, we can count the paths running n blocks north and n blocks east that do touch the dotted line by breaking them at the first touch and reflecting the second part of the path in the dotted line; in exactly the same way we see that *the number of paths running n blocks north and n blocks east that do touch the dotted line is equal to the number of all paths running $n-1$ blocks north and $n+1$ blocks east.*

And that's it: the Catalan number c_n—that is, the number of paths running n blocks north and n blocks east that stay on or above the diagonal—is the total number $\binom{2n}{n}$ of paths going n blocks north and n blocks east, minus the number $\binom{2n}{n-1}$ of such paths that touch the dotted line, so that we have a formula

$$c_n = \binom{2n}{n} - \binom{2n}{n-1}.$$

All that remains is to massage this answer a little, to take advantage of common factors in the two terms. Here we go: we've just seen that

$$c_n = \binom{2n}{n} - \binom{2n}{n-1}$$

and expressing these in terms of factorials we have

$$= \frac{(2n)!}{n!n!} - \frac{(2n)!}{(n-1)!(n+1)!}.$$

Clearing denominators gives us

$$= (n+1) \cdot \frac{(2n)!}{n!(n+1)!} - n \cdot \frac{(2n)!}{n!(n+1)!}$$

which we can combine to yield

$$= \frac{(2n)!}{n!(n+1)!}.$$

Writing the $(n+1)!$ in the denominator as $(n+1) \cdot n!$, we can express this as

$$= \frac{1}{n+1} \cdot \frac{(2n)!}{n!n!}$$

or in other words,

$$= \frac{1}{n+1}\binom{2n}{n},$$

and we're done. Our formula is proved, and more to the point the chapter is finished—or almost.

7.7 Why Do We Do These Things, Anyway?

Of what practical value is any of this? This is surely a question that's come up by now, and we (like most writers of math books) are all too aware of it. We're asking you to spend a not insubstantial amount of time and energy exploring a very esoteric world. What do you stand to gain by it? Why do we do these things, anyway?

There are, traditionally, a number of bogus answers to this question.

Bogus Answer #1: To calculate probabilities. In Section 5 we saw how we could use counting skills to estimate probabilities. Surely that's worth something?

Well, yes and no. It's true that counting, of the sort we've introduced here, is the cornerstone of probability theory. But in real life it has pretty limited utility, at least until you get into far more advanced techniques. Let's face it: when you're trying to decide whether or not to call the woman across the table who seems to be saying she has a full house to your flush, it can't hurt to know the odds—but the real question is usually, was that little twitch when she raised inadvertent or deliberate?

Bogus Answer #2: To aid in decision making. Of course knowing how many outfits you can make with four shirts and three pairs of pants doesn't help you actually decide what to wear. But in other situations—for example, distributing the widgets to the warehouses, as in the example at the beginning of this chapter—knowing how many possibilities there are can affect how you go about making the decision among them. It'll give you an idea, for example, of whether it'd be feasible to analyze each possible choice individually, or whether you need some other approach.

This may sound more far-fetched than the first, though it's at least as valid. Again, though, the utility is limited: only in relatively rare circumstances will this sort of calculation be of practical value.

Bogus Answer #3: Because mathematics always seems to have applications, even if you don't know what they are when you start out.

The best so far, actually. It really is amazing how even the most abstract mathematics seems to wind up having relevance to the real world far beyond what might have been anticipated. In fact, we'll see an absolutely spectacular example of this in Part IV of this book, if you stick around. But let's be real: if studying this stuff isn't at least a little bit rewarding to you in itself, the hope of some unspecified application sometime in the future doesn't really cut it as motivation.

The Real Answer: Because it's just plain fascinating. When you come right down to it, this is not stuff you study in the hopes of a concrete payoff; it's stuff you study just because it's so damn intriguing. Its beauty may not be as immediate as the beauty of music, or art, or literature—you have to dig a little deeper to find it—but it's there. And if you'll bear with us for the remainder of this text, we'll do our best to show it to you.

"A BUSHEL IS A UNIT OF WEIGHT EQUAL TO FOUR PECKS."
WHAT'S A PECK?
A QUICK SMOOCH.
© 1986 Universal Press Syndicate
YOU KNOW, I DON'T UNDERSTAND MATH AT ALL.
WATTERSON
2-3

PART II

Arithmetic

Chapter 8

Divisibility

We're now ready to begin our study of what is truly the heart and soul of mathematics: the arithmetic of whole numbers.

Mathematics is a funny world: very often the most profound insights are arrived at, not by confronting the outstanding problems of the day head-on, but by pursuing a seemingly minor question and seeing where it leads. It's sort of like walking around Rome looking for a coffee bar, turning a corner, and discovering the Piazza Navona. Mathematicians develop a sense of what sort of problems are apt to lead to this kind of discovery—what's a good neighborhood to wander around in, so to speak. In fact, it's probably more of a factor in what makes a mathematician successful than technical knowledge or computational speed.

The question we're going to start with here is about as minor as you can get: we want to ask, simply, *what numbers are divisible by both 4 and 6*? Think about it a moment before you start in on the chapter.

8.1 What Numbers *Are* Divisible by Both 4 and 6?

Well, we'll start with the obvious: $4 \times 6 = 24$ has to be divisible by both 4 and 6. And, of course, any multiple of 24 must be as well.

Are there others? To find out, we'll do the simplest thing possible: we'll make a list of multiples of 4 and a list of multiples of 6 and see what shows up on both lists.

4		8		12		16		20		24
	6			12			18			24

From this we find that 12 is another number divisible by both 4 and 6; and it follows that any multiple of 12 is as well.

Are there any others? Well, if we had an infinite amount of time and paper we could continue both lists forever, and see if anything else showed up; but that's not a practical solution to our problem. Instead, let's try a different approach: let's suppose someone hands us a number—to be concrete, say 392—and we're supposed to figure out if it's divisible by both 4 and 6. What we can do is to divide that number by 12 and get a remainder between 0 and 11. For example, if we start with 392, we find that 12 goes into 392 32 times, with a remainder of 8, or in other words,

$$392 = 32 \times 12 + 8.$$

Now, 8 isn't divisible by 6, and so we see that 392 can't be either: we know that 32×12 is divisible by 6, and if 392 were as well their difference $392 - (32 \times 12) = 8$ would be.

Obviously, we didn't need to do this just to see if 392 is divisible by 6; we could have just divided 392 by 6 and seen directly that the division didn't come out even. But this shows us a general approach to the problem. In general we can express any number n as a multiple of 12, plus a number between 0 and 11: that is, we can write

$$n = m \times 12 + r$$

with r a number between 0 and 11 (or, as we'd write it in math, $0 \leq r \leq 11$). There are then two possibilities:

- If the remainder r after division by 12 is 0, then the number n we started with is in fact a multiple of 12, and so is divisible by both 4 and 6; and on the other hand
- If the number n we started with is not divisible by 12, then remainder r is not 0—that is, r is between 1 and 11. Now, we see from our list that 12 is the smallest number divisible by both 4 and 6: no number between 1 and 11 is. So r isn't divisible by both 4 and 6; and since $12m$ *is* divisible by both 4 and 6, we see that the number n can't be.

To summarize: a number that isn't divisible by 12 can't be divisible by both 4 and 6, and we conclude that *the numbers divisible by both* 4 *and* 6 *are exactly the multiples of* 12.

We'll go on in a page or two and see how these ideas can be generalized; in fact, they'll form the cornerstone of this part of the book. But since we've just spent a lot of time counting things, we thought we'd take a moment out and do a counting problem involving this fact.

EXAMPLE 8.1.1 How many numbers between 134 and 387 are not divisible by either 4 or 6?

SOLUTION Let's consider how we might go about answering this (short of making a list of all the numbers, crossing off those divisible by 4 or 6, and counting what's left). The most reasonable approach would seem to be to use the subtraction principle. To start with, we know how many numbers between 134 and 387 there are altogether: by the basic formula of Section 1.1, it's just

$$387 - 134 + 1 = 254.$$

Next, we can figure out how many numbers between 134 and 387 are divisible by 4. Dividing by 4, we see that the smallest number in this range that's divisible by 4 is $136 = 4 \times 34$, and the largest is $384 = 4 \times 96$. So the numbers in this range that are divisible by 4 correspond to whole numbers between 34 and 96 inclusive, and there are

$$96 - 34 + 1 = 63$$

of them.

Similarly, we know how many of the numbers under consideration are divisible by 6: since the smallest number divisible by 6 in this range is $138 = 6 \times 23$ and the largest is $384 = 6 \times 64$, the numbers between 134 and 387 that are divisible by 6 correspond to whole numbers between 23 and 64 inclusive, so there are

$$64 - 23 + 1 = 42.$$

Naively, then, we want to start with the count of all numbers between 134 and 387—that is, 254—and subtract the number 63 of those divisible by 4, and the number 42 of those divisible by 6. But *we will then have subtracted twice those divisible by both 4 and 6*, and so we have to add those back in. This requires that we count the numbers between 134 and 387 that are divisible by both 4 and 6. Happily, we've just learned that the numbers divisible by both 4 and 6 are exactly the numbers divisible by 12, and so we can count these as well! The smallest number in this range divisible by 12 is $144 = 12 \times 12$, and the largest is $384 = 12 \times 32$, so there are

$$32 - 12 + 1 = 21$$

of them. The answer to our original question—how many numbers between 134 and 387 are divisible by neither 4 nor 6—is thus

$$254 - 63 - 42 + 21 = 170.$$

■

8.2 Least Common Multiples

Well, that certainly seemed like more trouble than it could possibly be worth to answer such a specialized question. But, as often happens in math, really understanding well one specific example of a problem leads us to a much broader answer. In this case the next step is to ask: what if we replaced 4 and 6 with two arbitrary numbers a and b? In other words, we pose the following:

Problem. Given two numbers a and b, say what numbers are divisible by both a and b.

In fact, the process we just went through tells us quite a bit in general. Suppose someone hands us a pair of numbers a and b—for another example, we could try 15 and 21—and asks us to describe the set of all numbers divisible by both. The first step would be find the smallest number that is divisible by both. This number is called the *least common multiple* of a and b, and is written $\text{lcm}(a, b)$ for short. In a pinch, we can always calculate the least common multiple of a and b by writing down multiples of a until we find the first one divisible by b, or vice versa. The only problem with this method is that it may take quite a bit of time. In the case of 15 and 21 we find the least common multiple is

$$\text{lcm}(15, 21) = 105 = 7 \times 15 = 5 \times 21.$$

Exercise 8.2.1 Find the least common multiple of the following pairs of numbers:

1. 2 and 5
2. 3 and 7
3. 9 and 15
4. 14 and 21

5. 12 and 22

6. 8 and 27

At this point, the same logic that we used in the case of 4 and 6 takes over. Suppose we've found the least common multiple of a and b. In the example we're currently following, $a = 15$, $b = 21$ and, as we said, $\mathrm{lcm}(15, 21) = 105$. Now, suppose we're asked whether a given number n—for example, say 882—is divisible by both 15 and 21. To answer this, we could simply divide 105 into 882 and calculate the remainder r: in this case, 105 goes into 876 eight times, so we write

$$882 = 8 \times 105 + 42.$$

The remainder in this case is not zero, and so it can't be divisible by both 15 and 21—remember that no number between 1 and 104 can be divisible by both. In this case, the remainder 42 is not divisible by 15. But now the remainder 42 differs from the original number 882 by a multiple of 105, which is a multiple of 15, and it follows that our original number 882 could not have been divisible by 15 to begin with.

Now, to answer the question in general, imagine that we're asked whether a given number n is divisible by both of two numbers a and b. Whatever n is, we can simply divide $\mathrm{lcm}(a, b)$ into n and write

$$n = m \cdot \mathrm{lcm}(a, b) + r$$

with the remainder r a number between 0 and $\mathrm{lcm}(a, b) - 1$. If the remainder $r = 0$, the number n is a multiple of $\mathrm{lcm}(a, b)$, and so it is certainly a multiple of both a and b. On the other hand, if the remainder is other than zero—as it was in the last example—then it can't be divisible by both a and b; and it follows that our original number n couldn't have been divisible by both to begin with. In conclusion, then, we have the answer to our original problem:

> The numbers divisible by both a and b are exactly the multiples of $\mathrm{lcm}(a, b)$.

Before we go on, take a moment to reflect on how we went about solving this problem. We first answered it for particular triples of numbers—first $a = 4$, $b = 6$ and $n = 392$, and then $a = 15$, $b = 21$ and $n = 882$—and then, when a pattern emerged, we made the leap of imagining how the same calculation would go for arbitrary numbers a, b and n. Of course, we had to find the right approach to the specific cases, or it might not have generalized as readily: if, in response to the question, "Is 882 divisible by both 15 and 21?" we had simply said, "Well, divide by both and see" it wouldn't have shed much light on the general case.

This is how much of mathematics is developed—we do a problem over and over in concrete cases, until we find an approach that we can carry out in general—and we'll use this method repeatedly. What's more, you should bear it in mind yourself whenever you're reading this book (or any math book, for that matter): if, in the course of deriving some formula or fact, we seem a little quick to haul out the algebra, you can always go back and try answering the question at hand for particular numbers, until the substitution of letters for numbers becomes clearer.

8.3 Greatest Common Divisors

At this point you may be moved to complain: "You promised to tell us how to find all numbers divisible by each of two given numbers a and b. But your answer involves a new quantity, $\text{lcm}(a, b)$. How are we supposed to find that?" Well, we sort of answered that question: you can just make a list of multiples of a and b until you find the first overlap. But you might complain that for large numbers a and b, finding $\text{lcm}(a, b)$ by this method is impractical.

You'd be right. We need to think more about least common multiples and how to calculate them. In particular, we'd like to focus on one question. For any two numbers a and b, the product ab is certainly divisible by both, and so—from what we've seen—the least common multiple $\text{lcm}(a, b)$ must be a number dividing the product ab. The question we'd like to ask now is, when is it equal to the product? More generally, since the product ab is a multiple of $\text{lcm}(a, b)$, we could ask: what is the ratio $ab/\text{lcm}(a, b)$?

As usual, when we don't know how to proceed, we can always work out some examples. That is, we can pick some pair of numbers a and b, work out their least common multiple, and compare it to the product ab. If we make a table, perhaps some pattern will present itself. We'll start with the examples we've worked out above and in Exercise 8.2.1, and throw in a few others:

TABLE 8-1 Some Least Common Multiples

a	b	ab	$\text{lcm}(a, b)$	$\frac{ab}{\text{lcm}(a,b)}$
2	5	10	10	1
4	6	24	12	2
3	7	21	21	1
4	9	36	36	1
9	15	135	45	3
10	21	210	210	1
14	21	294	42	7
15	21	315	105	3
12	22	264	132	2
8	27	216	216	1

There are a number of pairs (a, b) for which the least common multiple $\text{lcm}(a, b)$ is equal to the product ab, like the pairs (2,5), (3,7) and (4,9). There are others, like our examples (4,6) and (15,21) above, where it isn't. What distinguishes the two? Look closely at the table and see if you can figure out what's different about the pairs for which $\text{lcm}(a, b) = ab$, and those for which it isn't.

STOP.

CLOSE THE BOOK.

GRAB A PAD OF PAPER AND A PEN.

WORK OUT SOME EXAMPLES ON YOUR OWN.

THINK.

The answer is that whenever there's a number (other than 1) that divides both a and b—when they have a common divisor—then the least common multiple is less than the product; when there is no number other than 1 dividing both a and b, we see that $\text{lcm}(a, b) = ab$.

But we're not done. If we look closely at the table, something more emerges: the ratio $ab/\text{lcm}(a, b)$ is always a number dividing both a and b. In fact, we can check each of these cases and see that the ratio $ab/\text{lcm}(a, b)$ is the *largest* number dividing both a and b!

This calls for a definition. Given two numbers a and b, we will write $\gcd(a, b)$ for the *greatest common divisor* of a and b: that is, the largest number dividing both. The pattern that we see in the table can be expressed by the simple rule

For any two numbers a and b,

$$\text{lcm}(a, b) = \frac{ab}{\gcd(a, b)}.$$

We'll see why this is always true in Section 12.4 below.

8.4 Euclid's Algorithm

If the rule in the box above is true, and we will soon see that it is, we've reduced the problem of calculating $\text{lcm}(a, b)$ to that of calculating $\gcd(a, b)$. This only represents mathematical progress if we can come up with a simple method of calculating the greatest common divisor of two numbers. But here is where we have stumbled onto a great piece of mathematics, the *Euclidean algorithm*, one of the most efficient mathematical processes ever invented. (If you're not familiar with the term "algorithm," see Section 8.5 below.)

Let's start with an example. Suppose we want to find the greatest common divisor of two numbers—say, for instance, 30 and 69. (Of course these numbers are small enough that we could do this by hand, writing out the numbers that divide each and seeing how the two lists overlap; but we want to approach this more methodically.) The key to doing this, it turns out, is simplicity itself: we just divide one into the other, and find the remainder. In this case, 30 goes into 69 twice, with a remainder of 9; so we write

$$69 = 2 \times 30 + 9.$$

Now let's take a careful look at this equation, and what it means. Suppose first that a number c divides both 30 and 69. If it divides 30, it must divide $60 = 2 \times 30$, of course; and then we see that it must divide 9 as well, since 9 is the difference of 69 and 2×30. Thus, any number that divides both 30 and 69 must divide 9 as well.

We can also apply this logic in the other direction. Suppose we have a number that divides both 9 and 30. Again, it follows that it divides 2×30, and so it must divide 69 as well. In sum, we see that *the numbers that divide both* 30 *and* 69 *are exactly the numbers that divide both* 9 *and* 30; in particular,

$$\gcd(30, 69) = \gcd(9, 30).$$

The point here is that we've reduced the problem to one involving smaller numbers; and we can repeat the process to reduce the size further until the answer becomes obvious. To find the greatest common denominator of 9 and 30, for example, we just divide 9 into 30, writing

$$30 = 3 \times 9 + 3$$

and by the same logic deduce that

$$\gcd(9, 30) = \gcd(3, 9).$$

But now at the next step we see that 3 actually divides 9; and so the greatest common divisor of 3 and 9 is 3 itself, which represents the answer to our original problem:

$$\gcd(30, 69) = \gcd(9, 30) = \gcd(3, 9) = 3.$$

Let's indicate how this goes in general. Suppose we're given two numbers a and b, and we're asked to find their greatest common denominator. If a is the smaller of the two, we divide a into b and call the remainder r; that is, if a goes into b some number m of times, we write

$$b = m \times a + r$$

with r a number between 0 and $a - 1$. There are two cases. If $r = 0$—that is, if b is just a multiple of a—then the greatest common divisor of a and b is just a itself. If the remainder r is nonzero, we argue as follows:

- If a number divides both a and b, it must divide r as well; and
- If a number divides both r and a, it must divide b as well. Therefore
- The numbers that divide both a and b are exactly the numbers that divide both r and a, and in particular

$$\gcd(a, b) = \gcd(r, a).$$

Continuing, we simply repeat this process as many times as necessary to find the greatest common divisor of a and b.

We may formalize this process as

The Euclidean Algorithm

To find the greatest common divisor of two numbers a and b (with b larger than a),

- Divide a into b, and let r be the remainder.
- If $r = 0$, then we're done; a divides b and the greatest common divisor is a.
- If $r \neq 0$, we replace the pair of numbers (a, b) with the pair (r, a) and go back to the first step.

We can illustrate the algorithm with a flowchart:

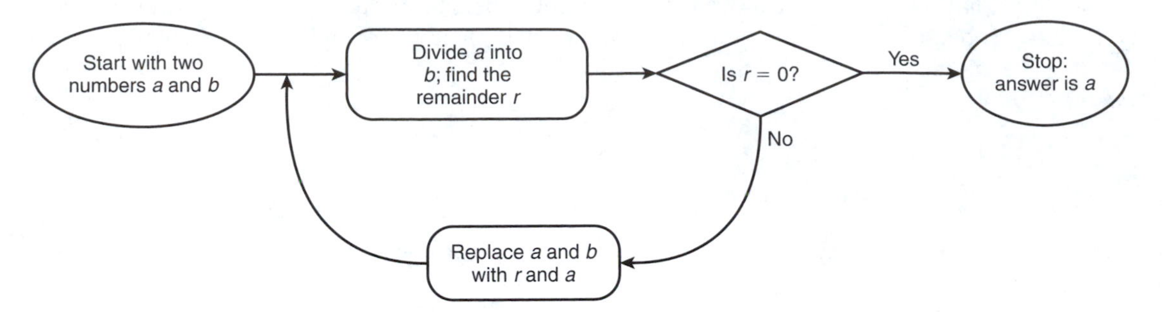

Let's see how this works in a sample calculation with larger numbers.

EXAMPLE 8.4.1 Find the greatest common divisor of 3,789 and 43,967.

SOLUTION We start by dividing 3,789 into 43,967. It goes 11 times, with a remainder of 2,288:

$$43{,}967 = 11 \times 3{,}789 + 2{,}288$$

so we replace the numbers 3,789 and 43,967 by the numbers 2,288 and 3,789. Now we divide again: 2,288 goes once into 3,789, with a remainder of 1,501:

$$3{,}789 = 1 \times 2{,}288 + 1{,}501.$$

Next, we divide 1,501 into 2,288:

$$2{,}288 = 1 \times 1{,}501 + 787;$$

we get a remainder of 787, so the next step is to divide that into 1,501:

$$1{,}501 = 1 \times 787 + 714.$$

Next, 714 goes into 787:

$$787 = 1 \times 714 + 73$$

and then we divide 73 into 714

$$714 = 9 \times 73 + 57$$

and 57 into 73:

$$73 = 1 \times 57 + 16.$$

We're getting close to the end: the last five steps are

$$\begin{aligned} 57 &= 3 \times 16 + 9 \\ 16 &= 1 \times 9 + 7 \\ 9 &= 1 \times 7 + 2 \\ 7 &= 3 \times 2 + 1 \end{aligned}$$

and finally

$$2 = 2 \times 1 + 0.$$

So the answer is that the greatest common divisor of 3,789 and 43,967 is 1—that is, there are no whole numbers bigger than 1 that divide both. ■

Now it's time for you to try your hand at the Euclidean algorithm.

Exercise 8.4.2 Find the greatest common divisor of the following pairs of numbers:

1. 39 and 72
2. 49 and 92
3. 216 and 594
4. 2,875 and 3,264

Of course, being able to find greatest common divisors allows us to find least common multiples, using the boxed rule at the end of Section 8.3.

Exercise 8.4.3 Find the least common multiple of each of the following pairs of numbers:

1. 39 and 65
2. 49 and 91
3. 144 and 216

As we said, there will be many more applications of the Euclidean algorithm throughout this part of the book—it's the mainspring that runs the clock. To close out this chapter, though, and just to see that we really have answered the question we set out to answer in the beginning, do the following.

Exercise 8.4.4 How many numbers are there between 600 and 6000 that are not divisible by either 39 or 65?

8.5 Who Was Euclid? What's an Algorithm?

Since we've just introduced something called the Euclidean algorithm, we should take a moment out and talk about who Euclid was, and what we mean by an algorithm in general.

To tell you the truth, we don't know the answer to the first question. In fact, we don't know if Euclid was a single mathematician, or the name chosen by a school of mathematicians in Platonic times. Whether singular or plural, Euclid compiled the results that appear in the famous text, *The Elements*. This is primarily a collection of the major results (prior to Archimedes) of Greek geometry, but it also contains a number of important results in number theory, such as the algorithm. It's one of the very few texts from that time that survived the middle ages.

If you're curious about the style or language of Euclid's book, we've included at the end of this chapter the title page and two sample pages (the ones where the algorithm is introduced) from the first English translation, published in London in 1570. Euclid clearly thought of his algorithm differently from us; for example, the quantities involved—the numbers we call a and b—are in Euclid's book lengths of line segments $\overline{AB}$ and $\overline{CD}$, and he speaks of one length "measuring" another, rather than one number dividing another evenly. But even though it's expressed in very different terms, you can see both the algorithm and the logic behind it clearly.

As for what an algorithm is, this is straightforward: an *algorithm* is a process designed to answer a question or solve a problem, with the properties that

- It proceeds in a fixed manner—without choices—from the original data; and
- It always terminates after a finite number of steps; that is, it can't run forever.

Actually, the word "algorithm" has an interesting history. It comes from the name of the Arab mathematician Al-Kwarismi, a scholar in the House of Wisdom in Baghdad during the rule of Harun al-Rashid in the 8th century. His famous text, *Hisab al-jabr w'al-muqabala*, gives us the modern term "algebra."

There are a couple of remarks to make about the Euclidean algorithm, the first of which is that it really is an algorithm in the sense above. It's clear that it satisfies the first condition—once you start with the two numbers a and b, the whole process is predetermined—so to see that it's an algorithm we just have to check that it always terminates after a finite number of steps. To see that this is the case, just observe that at each stage the remainder r is always strictly less than the number a you're dividing by, which is in turn the remainder at the previous stage. Thus, the remainders get consistently smaller and smaller; in the end, you must get a remainder of 0.

The other remark is that the process usually terminates pretty quickly. Think about it: if you take two numbers a and b at random, and divide a into b, the remainder could be any number between 0 and $a - 1$; on average, accordingly, it should be roughly half of a. That means that we might expect that each of the numbers we generate in the Euclidean algorithm would be on average roughly half the size of the one before it. There's no guarantee, obviously, but it suggests that if we start with a pair of numbers whose sizes are on the order of 2^n, the Euclidean algorithm should take around n steps to run. Thus, for example, if we start with numbers on the order of 1,000,000, which is very close to 2^{20}, we'd expect the Euclidean algorithm to take roughly 20 steps; if the numbers are on the order of a trillion (that is, 1,000,000,000,000, or a million million), 40 steps, and so on. Try it yourself with some randomly chosen pairs of numbers and see.

The most important thing here is that the number of steps in the Euclidean algorithm grows relatively slowly as the size of the numbers involved increases. If we're working with a pair of numbers of, say, 100 digits, the amount of time it would take a computer to find all the numbers that divide each and compare the lists would be many orders of magnitude larger than the lifetime of the universe. But the Euclidean algorithm will calculate their greatest common divisor in (roughly) 300 iterations—your desktop PC could do it in less than a second.

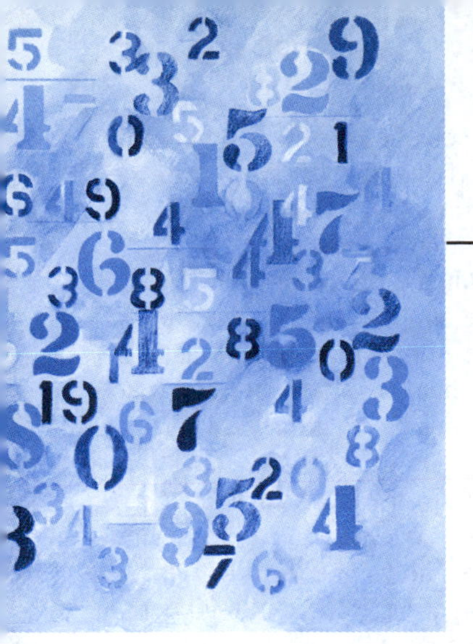

Chapter 9

Combinations

9.1 Pixels, Grackls, and Pancakes

Imagine you're playing a video game. You control a character—Gerry the Gerbil, let's say—who moves left and right. You can either hit the left or right key, which moves you a regular step of three pixels in either direction; or you can hold down the power key while you hit the left or right key, which moves you five pixels left or right.

Your goal right now is to pick up a pot of gold nearby. The problem is, to pick it up you have to land exactly on it, and right now you're just one pixel to the left of the pot. The question is: can you get the gold, and if so how?

OK, that's not a hard problem: any gamer would know immediately that you can move one pixel to the right by taking two regular steps right and one power step left. (Or you could take two power steps right and three regular steps left; there's more than one way. Try and find other ways!)

Now let's try making it a little more difficult: suppose now that a regular step is eight pixels left or right, and a power step is 13 pixels in either direction. Can we still do it?

This time we might have to think, but it's still possible. Look at it this way: if we can move 13 pixels one way and eight pixels the other, that means we can move five pixels in either direction. And if we can move eight pixels one way and five the other, we can move three pixels in either direction when we have to. And if we can move three pixels and five pixels, we already know how to do it. So it is possible; if you work it out, you see that five regular steps right and three power steps left will do it. (Again, there's more than one solution; we could also take five power steps right and eight regular steps left. Are there others?)

One more example of this problem: suppose a regular step is nine pixels in either direction, and a power step is 15. Can we still move one pixel to the right?

Well, we could try the same sort of logic as last time: if we can move nine and 15 left or right, we can move six in either direction. And if we can move six and nine, we can move three. But here we seem to be stuck: any way we move, the number of pixels we move is always going to be a multiple of three; and we're never going to get that gold.

Try some more examples of this, and then ask yourself the question in general: if a regular step is a pixels and a power step b pixels, for which a and b can we solve the problem? Why can we do it in the first two examples, but not the third?

We'll come back to this, but first let's shift gears a moment and. . . .

Imagine now that you find yourself in the small, isolated country of Fredonia. In Fredonia the basic unit of currency is the grackl. But the Minister of the Treasury is somewhat dotty and has issued two bizarre decrees. The first is that all transactions in Fredonia are to be carried out in cash; the second, that there will be only two denominations of coins: the trackl, worth three grackls, and the fackl, worth five.

Let's say you want to buy a newspaper, selling for one grackl. Assuming that both you and the newsdealer are adequately supplied with change, can you do it? You probably recognize this as a translation of the first version of the video game problem, and the answer, as it was there, is yes: for example, you could give the newsdealer two trackls, and receive a fackl in exchange; or you could give the newsdealer two fackls and receive three trackls in exchange.

OK, then, let's try it again: say the Minister has progressed from slightly dotty to full-bore loony, and has now decreed that a trackl is to be worth 11 grackls and the fackl worth 29. Can you still carry out a transaction of one grackl?

The answer is yes. First of all, by giving someone a fackl and getting two trackls back, you can transfer seven grackls. If you can transfer seven grackls and 11 grackls, you can transfer four. And if you can transfer four grackls and seven grackls, you can transfer three; and if you can transfer three and four, you can transfer one.

Exercise 9.1.1 How would you actually go about paying for your paper?

Time to move on again. . . .

Imagine now that you find yourself one morning in a strange kitchen. You decide to make pancakes—you know a great recipe, and the kitchen is stocked with plenty of all the ingredients you need. But there's a problem: your recipe calls for one cup of flour, and you can only find two measuring cups, one 5-cup measure and one 8-cup.

By now you know where we're headed, and you've probably already figured out the answer: to measure out one cup of flour you could put 16 cups of flour into the mixing bowl by filling the 8-cup measure twice, and then remove 15 cups by filling the 5-cup measure three times. Or you could add 25 cups by filling the 5-cup measure five times, and then take away three 8-cupfuls.

But now let's change the problem and suppose the two measuring cups hold, say, six cups and 10 cups. Can we still measure out one cup of flour? Well, we could measure out four cups easily by adding 10 and taking away six. And if we can measure four cups and six cups, we can measure two. But once more we're stuck: we can only measure even numbers of cups.

So, again, ask yourself the question in general: if a trackl is worth a grackls and a fackl is worth b, what transactions can we carry out? If our two measuring cups hold a cups and b cups, what quantities of flour can we measure out? We'll answer these questions (actually, this question: they're all the same question) in the next section.

9.2 Combinations

It's time to say mathematically what the last three problems are all about. We'll start with a definition: given two numbers a and b, a sum of a multiple of a and a multiple of b (allowing negative multiples) will be called a *combination* of a and b. In these terms, we can express the two questions we've dealt with repeatedly in the stories above as:

- Given two numbers a and b, what numbers n are combinations of a and b?; and
- If n is in fact a combination of a and b, how do we actually find multiples of a and b adding up to n?

To put it differently, if we're given numbers a, b and n, we're asking first whether or not we can solve the equation

$$x \cdot a + y \cdot b = n$$

in whole numbers (positive or negative) x and y; and secondly, if solutions do exist, how we can find them. Thus, for example, in the second grackl problem, we're asking for a pair of numbers x and y—positive or negative, but whole numbers—such that

$$x \cdot 11 + y \cdot 29 = 1.$$

A parenthetical note: It might strike you as splitting hairs to call this two problems. But to a mathematician, the question of whether a solution to a problem exists, and the question of how to find it, are two very distinct issues; and there are many occasions when we might want to study the first in its own right. There are a number of reasons for this. For one, we don't want to beat our heads against a wall trying to solve an insoluble problem. For another, very often an understanding of *why* a solution exists will actually lead us to a method for finding it. And finally, there are circumstances where we can determine that there is a solution to a particular problem, but it turns out to be impossible to find it explicitly. In those cases it just sounds better to say we solved one of two problems than to say we struck out.

But this isn't one of those cases; a few pages from now we'll have the complete solution. We do want to start, though, with the first question—that is, by asking when a number n can be expressed as a combination of two given numbers a and b.

To answer this, let's look first at the two examples we encountered a moment ago where we knew we couldn't find a solution. In the case of the character in the video game who could move either nine or 15 pixels in either direction, we observed that any motion involved a number of pixels divisible by three, and that therefore no combination of moves could result in anything but a number divisible by three. Similarly, in trying to measure out a cup of flour with measuring cups holding six cups and 10 cups, we saw that we could never measure an odd number of cups in that way.

The moral seems clear: if a number divides both a and b, it divides any combination of a and b. Now, the greatest common divisor $\gcd(a, b)$ divides both a and b, and any number that divides both divides $\gcd(a, b)$. So we have our first statement, albeit a negative one: *we can only solve the problem when the number n is a multiple of the greatest common divisor* $\gcd(a, b)$ *of a and b*.

The next question, clearly, is whether this condition is sufficient: whether it is in fact possible to solve the equation whenever n is a multiple of $\gcd(a, b)$. To answer this, let's go back and review the logic that we used to settle the issue affirmatively in the case of, say, the video game character who could move either eight or 13 pixels at a time in either direction. The way we argued that it's possible to move a single pixel—before we simply announced the solution—was to say:

- First, if we can move either eight or 13 pixels at a time in either direction, then we can certainly move the difference, or five pixels, in either direction.
- Second, if we can move either five or eight pixels at a time in either direction, then by the same logic we can move the difference, or three pixels, in either direction.
- Third, if we can move either three or five pixels at a time in either direction, it follows that we can move the difference, or two pixels, in either direction.
- Lastly, if we can move either two or three pixels at a time in either direction, then—finally—it follows that we can move the difference, or a single pixel, in either direction.

Now, stop reading right here. Grab a pad of paper and a pen, and do the following problem:

Exercise 9.2.1 Use the Euclidean algorithm to find the greatest common denominator of 8 and 13.

If you do the exercise, you'll see the point: *the logic we used to conclude that we could, in theory, solve the video game problem follows exactly the steps of the Euclidean algorithm applied to* 8 *and* 13.

Why is this? We'll look more closely now at this connection, but first a warning: there's a fairly high density of numbers and letters (of the algebra kind) on the following page. So get ready to hunker down; or, if that's the kind of mix that makes your eyeballs glaze over no matter how much coffee you've had, skip directly to the boxed statement below, go on and read the following two sections, and then come back to this discussion—it should be more transparent after you've had some practice.

OK, ready? Suppose now that we're given a pair of numbers a and b, with a less than b, and that we carry out one step of the Euclidean algorithm on them—that is, we divide a into b and find the remainder r, writing

$$b = m \cdot a + r$$

for some whole number m, and r between 0 and $a - 1$. The key point is that *the combinations of a and b are exactly the same as the combinations of r and a.*

Let's do an example to illustrate this. Suppose we start with the pair of numbers $a = 5$ and $b = 13$. Carrying out one step of the Euclidean algorithm, we write

$$13 = 2 \times 5 + 3.$$

Now, we can take any combination $x \cdot 5 + y \cdot 13$ and convert it into a combination of 3 and 5 by substituting $2 \times 5 + 3$ for 13:

$$\begin{aligned} x \cdot 5 + y \cdot 13 &= x \cdot 5 + y \cdot (2 \times 5 + 3) \\ &= y \cdot 3 + (x + 2y) \cdot 5. \end{aligned}$$

Going the other way, we can start with a combination of 3 and 5 and convert it into a combination of 5 and 13 by substituting $13 - 2 \times 5$ for 3:

$$\begin{aligned} x \cdot 3 + y \cdot 5 &= x \cdot (13 - 2 \times 5) + y \cdot 5 \\ &= (y - 2x) \cdot 5 + x \cdot 13. \end{aligned}$$

For example, we know how to express 1 as a combination of 3 and 5:

$$1 = 2 \times 3 + (-1) \times 5.$$

To express 1 as a combination of 5 and 13, then, we just write $13 - 2 \cdot 5$ for 3, and collect terms:

$$\begin{aligned} 1 &= 2 \times 3 + (-1) \times 5 \\ &= 2 \times (13 - 2 \times 5) + (-1) \times 5 \\ &= (-5) \times 5 + 2 \times 13 \end{aligned}$$

Now, suppose we start with two numbers a and b, and run the Euclidean algorithm: that is, we divide a into b, find the remainder r, replace the original pair (a, b) with the pair (r, a), and repeat the process. At each stage, we get in this way a new pair of numbers; and at the last stage of the algorithm, we arrive at a pair of numbers A and B with A dividing B evenly. We then know that A is the greatest common divisor of the pair of numbers (a, b) we started with.

What can this tell us about combinations? Well, to start with, we know the combinations of A and B: since A divides B the combinations of the two are just the multiples of A, that is, the multiples of $\gcd(a, b)$. But we've just seen that *the combinations of the pair of numbers we get at each stage of the process are the same as the combinations of the original two numbers a and b!*—in particular, the combinations of the two numbers a and b we started with are exactly the same as the combinations of the numbers A and B at the final stage, which is to say multiples of $A = \gcd(a, b)$. We've arrived at the final answer to the first of our two questions:

> The combinations of two numbers a and b are exactly the multiples of $\gcd(a, b)$.

Exercise 9.2.2 Rob and Sam are farmers who raise both sheep and chickens. Sheep are worth \$15, and chickens are worth \$9. Rob also has a frog worth \$12, which Sam wishes to have.

1. Rob and Sam agree in principle to make a fair trade, in the sense that Rob trades the frog together with some of his sheep and chickens to Sam, in exchange for livestock of equal value. Is this possible?
2. Now do the same problem assuming sheep are worth \$351 and chickens are worth \$30 (the frog is still worth \$12).

9.3 Finding Explicit Solutions

It's time now to deal with the second problem: in situations where the boxed statement tells us that a number is a combination of a and b, how do we actually find multiples of a and b adding up to that number?

The answer is suggested by the same logic that led us to the boxed statement in the first place. When we carry out one step of the Euclidean algorithm—that is, we start with two numbers a and b, write

$$b = m \cdot a + r$$

and replace the pair (a, b) with the pair (r, a)—we saw that the combinations of a and b are the same as the combinations of r and a. The point is, we didn't just assert that in the abstract; we actually saw how to convert a combination of r and a into a combination of a and b.

What this means is that, once we've run the Euclidean algorithm on a pair of numbers a and b and found their greatest common divisor, we can go back through the steps of the algorithm in reverse order, and at each step express $\gcd(a, b)$ as a combination of the pair of numbers we have at that step.

This is probably best shown by example. Let's try the following:

EXAMPLE 9.3.1 Express 1 as a combination of 271 and 670.

(We're not going to tell you to imagine yourself in a kitchen with measuring cups holding 271 cups and 670 cups or whatever, but if you need to, go right ahead.)

SOLUTION We first run the Euclidean algorithm forward to see if the problem is soluble at all: starting with the pair $(271, 670)$, we divide and write

$$670 = 2 \times 271 + 128;$$

so our second pair is $(128, 271)$. Next, we write

$$271 = 2 \times 128 + 15;$$

so our third pair is $(15, 128)$. Continuing, we have

$$128 = 8 \times 15 + 8;$$

so our next pair is $(8, 15)$; then

$$15 = 1 \times 8 + 7$$

and we've arrived at the pair $(7, 8)$. We're just about done now: in the last step,

$$8 = 1 \times 7 + 1,$$

and since 1 divides 7, were done: the greatest common divisor of 271 and 670 is 1.

Which means, in theory, that we should be able to express 1 as a combination of 271 and 670. But how?

The answer is that we start at the end of the Euclidean algorithm and work our way back up. To start with, 1 is obviously a combination of our last pair $(7, 8)$: we can write

$$1 = -7 + 8.$$

But now we've seen that *any combination of* 7 *and* 8 *is also a combination of our next-to-last pair* 8 *and* 15. All we have to do is to replace the "7" in the last expression by "$(15 - 8)$":

$$\begin{aligned} 1 &= -1 \times 7 + 1 \times 8 \\ &= -1 \times (15 - 8) + 1 \times 8 \\ &= 2 \times 8 - 1 \times 15. \end{aligned}$$

Once more: any combination of 8 and 15 is also a combination of 15 and 128, and we can carry out the conversion just by substituting $(128 - 8 \times 15)$ for 8:

$$\begin{aligned} 1 &= 2 \times 8 - 1 \times 15 \\ &= 2 \times (128 - 8 \times 15) - 1 \times 15 \\ &= -17 \times 15 + 2 \times 128. \end{aligned}$$

Next, any combination of 15 and 128 is a combination of 128 and 271:

$$\begin{aligned} 1 &= -17 \times 15 + 2 \times 128 \\ &= -17 \times (271 - 2 \times 128) + 2 \times 128 \\ &= 36 \times 128 - 17 \times 271 \end{aligned}$$

and finally, the very first step of the original Euclidean algorithm tells us that any combination of 128 and 271 is a combination of 271 and 670

$$\begin{aligned} 1 &= 36 \times 128 - 17 \times 271 \\ &= 36 \times (670 - 2 \times 271) - 17 \times 271 \\ &= -89 \times 271 + 36 \times 670. \end{aligned}$$

■

9.4 Any More Questions?

Q. *I think I got the idea, but could I try one more example to be sure?*

A. Absolutely. How about the problem we mentioned earlier—buying a newspaper for 1 grackl, when the trackl is worth 11 grackls and the fackl is worth 29. In other words, try to solve the equation

$$x \cdot 11 + y \cdot 29 = 1$$

with whole numbers x and y.

Q. *I think I can do that. You just run the Euclidean algorithm forward, writing*

$$\begin{aligned} 29 &= 2 \times 11 + 7 \\ 11 &= 1 \times 7 + 4 \\ 7 &= 1 \times 4 + 3 \\ 4 &= 1 \times 3 + 1. \end{aligned}$$

So at this point you know that there is a solution. Then you run the Euclidean Algorithm backwards: you write

$$1 = -1 \times 3 + 1 \times 4$$

(expressing 1 as a combination of 3 and 4)

$$\begin{aligned} &= -1 \times (7 - 4) + 1 \times 4 \\ &= 2 \times 4 + (-1) \times 7 \end{aligned}$$

(expressing 1 as a combination of 4 and 7)

$$\begin{aligned} &= 2 \times (11 - 7) + (-1) \times 7 \\ &= (-3) \times 7 + 2 \times 11 \end{aligned}$$

(expressing 1 as a combination of 7 and 11)

$$= (-3) \times (29 - 2 \times 11) + 2 \times 11$$
$$= 8 \times 11 - 3 \times 29.$$

and we're done; we've expressed 1 as a combination of 11 and 29. So it's really simplicity itself: you just hand the newsdealer eight trackls (worth 88 grackls), and he gives you three fackls (worth 87) in change. I don't know why more countries don't adopt the system.

But there's still one thing that bothers me. You said at one point that there would be more than one solution, and this process only seems to produce one. What happened?

A. Basically, once you have one solution you can generate others by using the fact that

$$11 \times 29 = 29 \times 11.$$

In terms of the currency example, this means you can trade 29 trackls for 11 fackls without altering the value of the transaction. For example, instead of giving the newsdealer eight trackls and getting three trackls in change, you could give him $8 + 29 = 37$ trackls, and get $3 + 11 = 14$ fackls in change. Or, going in the other direction, you could give him $8 - 29 = -21$ trackls, and receive $3 - 11 = -8$ fackls—in other words, you give him eight fackls, and he gives you 21 trackls and a paper. You should check that this all works, by the way.

Mathematically, what we're saying is that if x and y are one solution to the equation

$$x \cdot 11 + y \cdot 29 = 1,$$

then so is the pair $x + 29$ and $y - 11$; and likewise $x + 58$ and $y - 22$, and so on.

Q. You also said that we could solve the equation

$$x \cdot 11 + y \cdot 29 = n$$

whenever n is a multiple of the greatest common divisor of 11 *and* 29 *(which I guess is always, since* $\gcd(11, 29) = 1$*) and you haven't told me how to do that. What happens if I want to buy a steak that costs, say, seven grackls?*

A. You already figured out how to transfer one grackl; to transfer seven grackls, you can just multiply the quantities by seven. In the case of the one-grackl newspaper, as you worked it out, you just hand the newsdealer eight trackls, and he gives you the paper and three trackls in change. So if you wanted to buy a seven-grackl steak, you'd hand the butcher $7 \times 8 = 56$ trackls, and he'd give you the steak and $7 \times 3 = 21$ trackls in change.

Q. That's a lot of coins to carry around.

A. We'd have to agree, it's kind of cumbersome. But don't forget: once you've found a single solution, you can find others as we described a moment ago. So, for example, instead of giving the butcher 56 trackls, and getting 21 trackls in change, you could give the butcher $56 - 2 \times 29 = -2$ trackls and get back $21 - 2 \times 11 = -1$ fackls—in other words, you give the butcher one fackl and he gives you the steak and 2 trackls in change.

Q. Could you just sum up the whole process in general one more time?

A. Here you go:

How to Solve Linear Equations in Two Variables

Suppose you're given numbers a, b and n, and you're asked to express n as a combination of a and b—that is, solve the equation

$$x \cdot a + y \cdot b = n$$

in whole numbers x and y. Here's what you do:

- Run the Euclidean algorithm on a and b to find their greatest common divisor $\gcd(a, b)$.
- Check that n is a multiple of $\gcd(a, b)$; if it isn't, you can stop here because there is no solution to the problem.
- Now run the Euclidean algorithm in reverse to express $\gcd(a, b)$ as a combination of a and b: that is, solve the equation

  $$x \cdot a + y \cdot b = \gcd(a, b)$$

  in whole numbers x and y.
- Next, since n is a multiple of $\gcd(a, b)$, we can multiply this equation by the quotient $n/\gcd(a, b)$ to find a solution of the original problem.
- Finally, if you want to find other solutions, you can take the solution you found and add or subtract multiples of the equation

  $$b \cdot a - a \cdot b = 0.$$

Any more questions?

Q. Just one. Why are these called "linear equations" anyway?

A. Well, the "linear" part is because the numbers x and y you're trying to solve for appear in the equation only as multiples of fixed numbers—they're not squared, or multiplied by each other or anything. By way of contrast, an equation like $3x^2 + y^2 = 4$ would be called *quadratic*, or *second degree*; and an equation like $5x^3 + 7x^2 y = 13$ would be called *cubic*, or *third degree*.

In the mathematical literature, this is actually called a "linear Diophantine equation." The word "Diophantine" refers to the classical mathematician Diophantus of Alexandria, who lived in the 3rd century A.D. and whose book, *The Arithmetica*, is the first treatise on algebra and number theory. He was interested in finding whole-number solutions to equations of the first, second, and third degree, and several of his problems refer to the sort of linear equations we've just learned to solve.

Now it's time for you to get some practice.

Exercise 9.4.1

1. Find the greatest common divisor of 16 and 55.
2. Solve the equation $16x + 55y = 3$ in whole numbers x and y.

Exercise 9.4.2 Which of the following equations have solutions in whole numbers x and y? In each case, find a solution or say why there is none.

$$105 \cdot x + 24 \cdot y = 3$$

$$105 \cdot x + 24 \cdot y = 4$$

$$105 \cdot x + 24 \cdot y = 9$$

Exercise 9.4.3 Going back to the video game example, suppose your character can move either 14 pixels or 19 pixels in either direction. How can you move him one pixel right? How can you move him three pixels right?

Exercise 9.4.4 Solve the equation

$$x \cdot 22 + y \cdot 28 = 4$$

with whole numbers x and y.

Chapter 10

Primes

We said at the outset of this part of the book that the subject we're studying—the arithmetic of whole numbers—is the heart and soul of mathematics. The particular topic we're embarking on in this chapter—prime numbers and their role in arithmetic—is in turn at the core of arithmetic.

In the preceding two chapters we've developed a basic tool for studying whole numbers (or *integers*, as mathematicians call them): the Euclidean algorithm. We've already seen some applications of this tool, but in this chapter and the next we'll see how to use it to illuminate the basic nature of numbers themselves.

10.1 What Is a Prime?

Probably everybody knows this one: a *prime* is a number larger than 1 that has no divisors other than itself and 1. (The number 1 is not considered a prime by convention. The reason for this will be more apparent later in this chapter.)

Why are prime numbers so important? Well, imagine for the moment that you had at your disposal only the operation of addition; say multiplication didn't exist. If you just started with the number 1, you could build up the entire system of whole numbers just by adding 1 to itself over and over. But now suppose the opposite situation: say you could multiply, but not add. Now in order to generate all numbers, you would need to start with all the primes: a prime number, by definition, can't be obtained by multiplying other numbers. Thus, prime numbers are the fundamental building blocks of multiplication in the same way that the number 1 is the basic building block of addition.

Let's take a look at the first few primes. We'll start by writing out all the numbers from 2 on up:

2 3 4 5 6 7 8 9 10 11 12 13 14 15 16 17 18 19...

and consider them one by one:

2 is a prime
3 is a prime
  4 isn't a prime: it's equal to 2×2

5 is a prime
  6 isn't a prime; it's equal to 2×3
7 is a prime
  8 isn't a prime; it's equal to $2 \times 2 \times 2$
  9 isn't a prime; it's equal to 3×3
  10 isn't a prime; it's equal to 2×5
11 is a prime
  12 isn't a prime; it's equal to $2 \times 2 \times 3$
13 is a prime
  14 isn't a prime; it's equal to 2×7
  15 isn't a prime; it's equal to 3×5
  16 isn't a prime; it's equal to $2 \times 2 \times 2 \times 2$
17 is a prime
  18 isn't a prime; it's equal to $2 \times 3 \times 3$; and
19 is a prime.

So we see that the first few primes are

2 3 5 7 11 13 17 and 19.

We can extend this list a little further:

2 3 5 7 11 13 17 19 23 29 31 37 41 43 47 53 59...

See if you can list all the prime numbers less than 100.

10.2 Prime Factorization

The patterns in the list of primes are endlessly fascinating; we'll talk quite a bit more about them in this chapter and in later ones. But first we want to focus on the nonprime numbers in the list above. When we were describing the primes as building blocks of multiplication, we pointed out that you needed to include the primes in order to generate all numbers, since the primes themselves could not be generated any other way. But we dodged the issue of whether, conversely, the primes suffice to generate all numbers: is it the case that every number is a product of primes?

The answer, at least on the basis of the evidence in the list so far, is yes: every nonprime (or *composite*) number we encountered can be expressed as a product of primes. And it's not hard to see that this is the case in general. Suppose we start with any whole number—say, for example, 165. We can simply ask, "Is it divisible by 2?" "Is it divisible by 3?" "Is it divisible by 4?" and so on until we reach the number itself. If the answer each time is no, then the number is prime. If the answer at any point is "yes," that means the number we started with is a product of two smaller numbers, and now we can just start this process over for these two numbers.

In our example, 165 is not divisible by 2, but it is divisible by 3: we can write

$$165 = 3 \times 55.$$

Now we ask the same series of question about each of the factors 3 and 55. In the case of 3, we know it's prime; in the case of 55, when we get to the question "Is it divisible by 5?" the answer is yes; we have

$$55 = 5 \times 11$$

and these numbers are prime. Thus we have

$$165 = 3 \times 5 \times 11$$

so 165 is indeed a product of primes. In this way we can break down any whole number into a product of prime factors; this is called a *prime factorization* of the number.

Exercise 10.2.1 Find a prime factorization of the following numbers:

1. 85
2. 342
3. 851
4. 137

10.3 The Sieve of Eratosthenes

Let's go back now to the way we generated the list of the primes up to 19. This is an example of a basic technique for finding primes known to the ancient Greeks, called the *sieve of Eratosthenes*. Maybe you've heard of the classic technique for creating a sculpture of a horse: you start with a block of stone and chip away anything that doesn't look like a horse. The sieve of Eratosthenes works the same way: you start with a string of ordinary numbers, and systematically eliminate the composite, or nonprime, ones.

For example, suppose you want to find the primes in the string of numbers

50 51 52 53 54 55 56 57 58 59 60.

You start by eliminating the ones divisible by 2: that is, you cross out every other number, starting with 50.

~~50~~ 51 ~~52~~ 53 ~~54~~ 55 ~~56~~ 57 ~~58~~ 59 ~~60~~

Next, we eliminate the numbers divisible by 3. We start with 51, which is 3×17, and then we cross out every third number after that: 54 (which was eliminated already; now it's doubly out), 57 and 60.

~~50~~ ~~51~~ ~~52~~ 53 ~~54~~ 55 ~~56~~ ~~57~~ ~~58~~ 59 ~~60~~

Now, do we have to cross off the numbers divisible by 4? No, because a number divisible by 4 would have to be divisible by 2, and we've already gotten rid of those. In general, when we carry out the sieve of Eratosthenes, we only have to eliminate numbers divisible by primes, since every number that's not a prime will be divisible by one.

So on to 5: 50, which was already knocked out, gets knocked out again; 55 gets knocked out, and 60 gets knocked out for the third time. And after that comes 7, which eliminates 56 for the second time; we have at this point

~~50~~ ~~51~~ ~~52~~ 53 ~~54~~ ~~55~~ ~~56~~ ~~57~~ ~~58~~ 59 ~~60~~

You're probably getting a little tired of this by now, especially if you're anticipating having to carry out this process for 11, 13, 17 and all the rest of the primes less than 60. Well, we have good news: we can stop now. The point is,

> If a number n is not prime, then it must be divisible by a prime less than or equal to $\sqrt{n}$.

Think about it: if the number n is not prime, it can be written as a product of two smaller numbers

$$n = a \times b.$$

Now, the numbers a and b can't both be bigger than $\sqrt{n}$: if they were, the product would be bigger than n. So either a or b must be less than $\sqrt{n}$. Whichever it is, it's either prime itself or divisible by a prime less than itself, which is to say less than $\sqrt{n}$.

Thus, for example, if any number on our list were not a prime, it'd be divisible by a prime number less than $\sqrt{60}$. Now, we've already eliminated all the numbers on our original list divisible by the primes 2, 3, 5 and 7. The next prime is 11, and since $11^2 = 121$ is larger than 60, 11 itself must be larger than $\sqrt{60}$; so any nonprime number less than 60 must be divisible by 2, 3, 5 or 7. It follows that we're done: we've eliminated all the composite numbers on the list, and we can conclude that the numbers that are left—53 and 59—are prime.

To put it another way, the numbers on our list that are divisible by primes larger than 7—for example, 55, which is 5×11, or 57, which is 3×19—must also be divisible by primes less than 7; and so will have been eliminated already long before we come to 11 or 19.

In sum, when you're carrying out the sieve of Eratosthenes on a sequence of numbers, *you only have to eliminate the numbers divisible by primes less than the square root of the largest number in your sequence.*

Let's try this once again, say with the numbers between 210 and 220. We'll start this time by figuring out what primes we have to check. We see that

$$13^2 = 169 < 220$$

but

$$17^2 = 289 > 220$$

so $\sqrt{220}$ falls somewhere between 13 and 17; thus, for this sequence of numbers we only have to check divisibility by primes up to and including 13. We'll start by crossing off the even numbers:

~~210~~ 211 ~~212~~ 213 ~~214~~ 215 ~~216~~ 217 ~~218~~ 219 ~~220~~

Next, we do numbers divisible by 3, which kills 213 and 219 (as well as eliminating 210 and 216 a second time each):

~~210~~ 211 ~~212~~ ~~213~~ ~~214~~ 215 ~~216~~ 217 ~~218~~ ~~219~~ ~~220~~

Now we cross off numbers divisible by 5, which nails 215:

~~210~~ 211 ~~212~~ ~~213~~ ~~214~~ ~~215~~ ~~216~~ 217 ~~218~~ ~~219~~ ~~220~~

and 7 kills 217 (as well as 210 a fourth time!)

~~210~~ 211 ~~212~~ ~~213~~ ~~214~~ ~~215~~ ~~216~~ ~~217~~ ~~218~~ ~~219~~ ~~220~~

Finally, 11 and 13 don't do much except further eliminate numbers already crossed off (220 and 218, respectively). So our conclusion is, the only prime number between 210 and 220 is 211.

Exercise 10.3.1 Carry out the sieve of Eratosthenes to find the primes between 400 and 420.

10.4 Some Questions

Already this suggests a number of questions we might ask about prime numbers. To begin with, is there any way to predict how many primes there'll be in a given sequence like this? Is there always a prime in any sequence of 10 numbers? (The answer is no; check the sequence 200-210.) How about in a sequence of 20, or 100, or 1000 numbers in a row—will there always be a prime among them?

At the other end, we could ask about "adjacent pairs" of prime numbers. If we take two numbers in sequence, one of the two must be even, so they can't both be prime (unless they're 2 and 3, of course). But what about pairs of numbers of the form n and $n + 2$—that is, a number and the number 2 greater. Can they both be prime? Of course they can: look at 5 and 7, or 17 and 19, or 29 and 31, or 41 and 43, or.... Such pairs of primes are called *twin primes*, and the ellipsis at the end of the last sentence suggests a question: how many pairs of twin primes can we find?

For that matter, how frequent are the primes themselves? Since there are infinitely many numbers, it doesn't make sense to ask what fraction of all numbers are prime. But there is a way of making a sort of sense of this question. To begin with, we can certainly ask what proportion of all numbers between 1 and, say, 1,000,000 are prime—that is, if we generate a six-digit number randomly, for example by rolling a 10-sided die six times, what are the odds that it will be prime? Then we can ask, among all 10 digit numbers, or 20 digit numbers, what fraction are prime, and so on. Finally, we can ask how this fraction behaves as the number of digits in our numbers increases: do the odds of finding a prime increase or decrease, or do they approach a fixed value?

There is one fundamental question that we need to address first: are there only a finite number of prime numbers, or are there infinitely many primes? Does the sequence of primes go on forever, in other words, or does it stop somewhere? Is there a "largest prime" of all—and if so, we might ask, what is it? This is the question we'll take up in the following section.

10.5 How Many Primes Are There?

The answer is that there are infinitely many primes: the list of primes goes on forever, though (as we'll see) they become sparser as the numbers get larger. The fact that there are an infinite number of primes was established in Euclid's *Elements*, and represents one of the triumphs of classical arithmetic.

It is not immediately obvious that this is the case. Nor is it possible—and this is a point that we really want to explain—to demonstrate directly that there are infinitely many primes. Instead, to convince you that there are infinitely many primes, we have to give an essentially indirect argument.

Put it this way: by far the simplest, most direct and probably most convincing way of showing you that there are infinitely many primes would be to exhibit a sequence of numbers, going on forever, all of which were prime. For example, suppose we want to convince you that there are infinitely many odd numbers. We could exhibit the sequence

$$1 \quad 3 \quad 5 \quad 7 \quad 9 \quad 11 \quad 13\ldots$$

whose n^{th} member is $2n-1$: this is an infinite sequence without repetitions, and every number in the sequence is odd; therefore, there exist infinitely many odd numbers. (The fact that our sequence actually includes *all* odd numbers is nice but logically irrelevant: to prove that there are infinitely many odd numbers we could just as well have exhibited the sequence

$$1 \quad 5 \quad 9 \quad 13 \quad 17 \quad 21 \quad 25\ldots$$

whose n^{th} member is $a_n = 4n - 3$.)

Or another way to argue the infinitude of odd numbers would be to show that, for any number n, there is an odd number greater than n—for example, $2n + 1$. This means there can't be a "largest odd number"—there's always a bigger one than the one you have, so the collection of odd numbers must be infinite.

But there is no analogously direct way of proving the infinitude of primes. That is, we do not know a simple formula that will generate an infinite string of prime numbers; nor is there an algorithm, given a number n, that will produce an explicit prime number larger than n. Instead, we will have to argue by contradiction.

Now, the idea of proof by contradiction is itself not unreasonable. Suppose you want to prove a statement A. You assume initially that A is false, and produce consequences of that assumption. If at some point you derive as a consequence a statement that you *know* to be false, you may conclude that your original hypothesis must have been erroneous. In other words, A must have been true.

But, as any mathematician who has ever taught a proof-based course knows, introducing proof by contradiction as a mode of argument—with all the warnings in the world about its potential misapplication—is something you do only with the greatest trepidation. The problem is this: when you've set up the problem so that the desired result is a contradiction, it's all too easy to introduce extraneous mistakes—either other, hidden hypotheses that may be false, or flaws in the logic—that yield such a contradiction, and so (seemingly) give the answer you want. In effect, you're setting it up so that a mistake will be rewarded by apparent success. Given the number of mistakes people make in normal circumstances, this is probably not a good idea.

OK; you've been warned. Let's put our trepidation aside and get down to it:

Assume that there are only finitely many prime numbers. Let's say there are n of them, for some number n, and give them names: $p_1, p_2, p_3, \ldots, p_n$. By what we've seen, we know that *every number bigger than 1 has a prime factorization, and so must be divisible by at least one of these primes.*

Now take all the primes $p_1, p_2, p_3, \ldots, p_n$ and multiply them together, add 1 to the result, and call this number m:

$$m = (p_1 \times p_2 \times \cdots \times p_n) + 1.$$

The first thing we see is that m can't be divisible by p_1: since the product $(p_1 \times p_2 \times \cdots \times p_n)$ is a multiple of p_1, m leaves the remainder 1 after division by p_1. By the same token, m can't be divisible by p_2, or p_3, or any prime on our list: it's 1 more than a multiple of each of them. But, as we said, every number must be divisible by at least one of the numbers $p_1, p_2, p_3, \ldots, p_n$. This is the desired contradiction, and we can conclude that the original assumption that there were only finitely many primes must have been erroneous.

Another way to phrase this argument would be the following. Suppose we have any finite collection of primes, again called $p_1, p_2, p_3, \ldots, p_n$. We claim that there must be at least one prime not on the list. How do we see this? Again, we multiply together all the primes on our list and add 1. It follows then that none of the primes $p_1, p_2, p_3, \ldots, p_n$ divide the resulting number m; or in other words, *every prime factor of m is a prime not on our list*—either m itself is a prime not on our list, or it's a product of primes not on our list. Thus, no matter how long a finite list of primes is, it can't be complete.

The argument is thus a constructive one in this sense: given any finite list of primes, we find one that's not on it. But we still don't know how to generate a list of *all* prime numbers; and in particular, even if we have succeeded in constructing a list of all the primes up to a certain point, we don't have any way (short of hauling out Eratosthenes' sieve) of finding the next one.

10.6 Prime Deserts and Twin Primes

As we said, there is a huge number of questions about the distribution of primes, most of which are subjects of mathematical inquiry to this day. We'll finish this chapter by talking a little more about two of them.

The first is one of the questions that arose naturally from the sieve of Eratosthenes. We asked: is it always the case that every string of 10 numbers in a row includes at least one prime? If not, does every string of 20? and so on.

What we will show you here is that no matter how large a number n you choose, there is a string of n numbers in a row with no primes. For example, suppose you want to find a sequence of, say, 35 numbers in a row with no primes. Consider the string

$$36!+2 \quad 36!+3 \quad 36!+4 \quad 36!+5 \quad \ldots \quad 36!+35 \quad 36!+36.$$

We claim that none of these numbers can be prime.

Why not? Well, to start with, observe that 36! is divisible by all the numbers from 2 up to 36—it's just their product, after all. It follows that

$36! + 2$ must be divisible by 2, since it's a sum of numbers divisible by 2
$36! + 3$ must be divisible by 3, since it's a sum of numbers divisible by 3
$36! + 4$ must be divisible by 4, since it's a sum of numbers divisible by 4

and so on, until finally we see that

$$36! + 36 \text{ must be divisible by } 36.$$

Note that we can't extend this in either direction: $36! + 1$ might well be prime—at any rate, we can't immediately point to any number that obviously divides it—and so might $36! + 37$. But starting with 36! and adding any number between 2 and 36 does give us a composite number, so we have our 35 numbers in a row without a prime.

You can see how this works in general: if we want 1,000 numbers in a row without a prime, for example, we can take the numbers

$$1{,}001! + 2, \quad 1{,}001! + 3, \quad 1{,}001! + 4, \ldots, 1{,}001! + 1{,}001.$$

What we see is that, as we go out through the list of all numbers, these *prime deserts*—stretches of numbers between primes—can and do become longer and longer.

But not consistently so! At the opposite extreme from prime deserts we have *twin primes*, which are pairs of primes of the form p and $p + 2$—that is, pairs of primes separated by the minimal possible space along the number line. And these also seem to go on forever: it is a classical and famous conjecture in mathematics—called, reasonably enough, the *twin prime conjecture*—that there are infinitely many twin primes. The truth of this conjecture remains unknown, though people have spent an inordinate amount of time looking. For example, it's known[1] that there are

35 twin prime pairs less than 1,000;
8,169 twin prime pairs less than 1,000,000;
3,424,506 twin prime pairs less than 1,000,000,000;
1,870,585,220 twin prime pairs less than 1,000,000,000,000; and
1,177,209,242,304 twin prime pairs less than 1,000,000,000,000,000.

As of 2001, the two largest known twin primes are

$$318{,}032{,}361 \times 2^{107{,}001} \pm 1$$

—numbers with 32,220 digits.[2] But we still don't know if there are an infinite number of them. As outlandish it may seem, we can't prove that they don't just keep recurring up through numbers with, say, 50,000 digits, and then stop.

In sum, then, the space between two successive primes can be arbitrarily large; or it can be 2. We just don't know, given one prime number p, where the next one is going to turn up.

[1] see for example the websites `http://www.mscs.dal.ca/~ joerg/res/tp-en.html` or `http://www.trnicely.net/counts.html`

[2] see `http://www.utm.edu/research/primes/largest.html`

Chapter 11

Factorization

11.1 A Note About the Exercises

Imagine for a moment that you find yourself in a hotel room in a strange city. You're going to be spending some time there, and you want to get to know the place, so you set out to wander. It's a little daunting: every street is unfamiliar, and branches out in turn into other unfamiliar streets. There's no way to decide which way to go, so you wander around at random, always wondering what you might have encountered if you had gone the other way at that last intersection.

But gradually things change. The third or fourth time out, you find yourself recognizing a landmark here or there. You find yourself at an intersection, and realize that you've already explored several of the roads leading out. The cafes and shops within a block or two of the hotel become familiar, and even when you venture further out you discover you have a sense of where you are. Even the language, over time, comes to sound less foreign. Eventually, if you stick around, it becomes your neighborhood, in your city.

Math is like that. To anyone exploring a new subject in math, it's an unfamiliar landscape and an unfamiliar language. (This is as true of grizzled professional mathematicians as it is of a beginning student. A mathematician may have the advantage of having already experienced the process of travelling to other foreign cities and knowing in general terms what to expect, but a new topic is still terra incognita.) And really, the only way to change that—to make it your neighborhood, in your city—is to get out there and explore. It may be disconcerting at first to wander around, not quite sure where you are or where you should be heading. But it's the only way: you can no more learn math just by reading a book than you can learn your new city just by sitting in your hotel room studying a street map.

The exercises scattered throughout this text are a way of getting out there and exploring, of getting familiar with the terrain. Of course, if you feel inclined to explore on your own as well, that's even better, but please *do the exercises*.

Remember: doing math is, more than anything else, about familiarity. The process by which a mathematician proves a theorem has little to do with the standard movie image of a disheveled, hypercaffeinated genius experiencing a moment of brilliant insight. It has a lot more to do with the way everyone knows the best way to get around their home town.

11.2 What a Prime Is, Really

Let's start with a simple question:

When is a product of two numbers even?

You probably know how to answer this: at some point in your life you've observed that a product of an even number with any number is even, while a product of two odd numbers is odd. The answer, accordingly, is that

A product ab is even precisely when either a or b is.

Now let's ask the same question, replacing "2" by "3":

When is a product of two numbers divisible by 3?

Well, half of the answer to the previous version applies as well in this case: if you multiply two numbers, and at least one of the factors is divisible by 3, then surely the product is. But what about the other half: if you multiply two numbers and the result is a multiple of 3, is it necessarily the case that one of the two you started with was a multiple of 3? To phrase it differently, can you find two numbers, neither divisible by 3, whose product is divisible by 3?

That's actually a tough question to answer with your bare hands, but try anyway: maybe think of some pairs of numbers not divisible by 3, and see in each case if the product is. Or else start with some numbers that are divisible by 3, and look at different ways of factoring them: is it always the case that one factor or the other is a multiple of 3? Since we wouldn't want you to feel that we're leaving all the heavy lifting to you, we'll do our bit: here's a multiplication table for numbers not divisible by 3, up to 8:

Table 11-1 Multiplication Table for Numbers Not Divisible by 3

×	1	2	4	5	7	8
1	1	2	4	5	7	8
2	2	4	8	10	14	16
4	4	8	16	20	28	32
5	5	10	20	25	35	40
7	7	14	28	35	49	56
8	8	16	32	40	56	64

None of the elements in the table are divisible by 3; this strongly suggests the answer to our question is that *the number* 3 *divides a product* $a \times b$ *only if it divides* a *or* b.

We could also replace the "3" above with 4, or 5, or any other number, ask the analogous question and carry out a similar experiment. If we do, we see pretty quickly that the corresponding statement is not always true. For example, suppose we replace "3" by "4," and ask if there are pairs of numbers a and b, neither divisible by 4, whose product is a multiple of 4. After a little playing around, we see that there are: for instance, neither 6 nor 10 is divisible by 4, but their product 60 is.

Exercise 11.2.1 Can you find a pair of numbers, neither divisible by 5, whose product is divisible by 5? Try some examples and see if you can find one; if not, can you find a reason why you can't? How about 6?

Let's get back now to the problem about numbers divisible by 3. The evidence we've got is clear: a product of two numbers not divisible by 3 can't be divisible by 3, at least not as far as this table goes. But does that mean it can never happen? How could we know that?

In answer, we are going to give an argument, using the Euclidean algorithm, to convince you of exactly that. We'll go more slowly than is perhaps necessary, and highlight each step, because ultimately we want to apply the same logic more generally to other numbers besides 3. (In particular, you should keep an eye out as we go through this argument, and see what special properties of the number 3 are invoked.)

- Start with two numbers—call them a and b—and suppose that their product $a \cdot b$ is divisible by 3.
- Suppose moreover that 3 does not divide a.
- The only numbers that divide 3 are 1 and 3 itself; and since 3 doesn't divide a it follows that

$$\gcd(3, a) = 1.$$

- It follows in turn that we can express 1 as a combination of 3 and a: that is, there are whole numbers x and y such that

$$x \cdot 3 + y \cdot a = 1.$$

- Next, multiply that last equation by b:

$$x \cdot 3 \cdot b + y \cdot a \cdot b = b.$$

- Now look at this last equation. The first term on the left is visibly a multiple of 3. The second term on the left is also a multiple of 3, because the product ab is! Thus their sum must likewise be a multiple of 3; in other words, we may conclude that *b must be divisible by* 3.

What have we shown by this argument? To recapitulate, we've shown that if 3 divides a product ab and doesn't divide a, then it must divide b. More symmetrically, we've shown that *if 3 divides ab it must divide either a or b.*

Now comes the crucial part: what's so special about the number 3? Where in this argument did we use any special properties of 3, and where would we go wrong if we replaced 3 throughout by 4, or 5, or 6?

The answer, if you go back over the argument step by step, is that we used a special property of the number 3 in only one place: the third step. There we said "the only numbers that divide 3 are 1 and 3 itself; and since 3 doesn't divide a it follows that $\gcd(3, a) = 1$." This wouldn't work if we replaced 3 by 4 or 6—a number a that isn't divisible by 6, for example, could very well have greatest common divisor $\gcd(6, a)$ bigger than 1, if a were even or divisible by 3. But *it does work if we replace 3 by any prime number* p: we can conclude that if a product ab of two numbers is a multiple of a prime number p, then one of the two factors a or b must have been a multiple of p to begin with.

In fact, the same logic applies to a product of any number of factors. If we had three numbers a, b and c, for example, none of which were divisible by a prime number p, then their product couldn't be either: if p didn't divide a or b, it couldn't divide their product ab; and if p didn't divide ab or c, it couldn't divide abc. In conclusion, the argument we've just gone through tells us a fundamental fact, one that merits a box of its own:

> If a prime number p divides a product of numbers then it must divide one of the factors.

Notice, finally, that this statement is false for any composite number: if a number n is a product of two smaller numbers a and b, then obviously n doesn't divide either a or b, but does divide their product; and we can use this to make up lots of other examples. For example, $6 = 2 \times 3$, which is an example right there of two numbers, neither divisible by 6, whose product is; for another example, we could take $10 = 2 \times 5$ and $21 = 3 \times 7$.

In fact, to a mathematician, it's the property in the box above, rather than the condition of not being itself a product of other numbers, that *defines* what it means to be a prime. As we've just seen, for ordinary numbers the two conditions are equivalent, so it's just splitting hairs which one we adopt as a definition. But in other number systems—about which we'll say more in Part III—the two notions don't necessarily coincide; and there the property in the box is taken as the definition of a prime.

There is another consequence of the statement in the box that we should mention. Suppose we have any whole number n, and a number b that divides it; and suppose a is the quotient—in other words, suppose we have an identity

$$a = \frac{n}{b}$$

with a, b, and n all whole numbers. Then we also have

$$n = a \cdot b.$$

Now, suppose p is any prime number that divides n. By what we've just seen, it must divide either a or b; by the same token, if p divides n and does not divide b, it must divide a. In English:

> If a prime number divides the numerator of a fraction but not the denominator, it must divide the quotient.

Again, notice this is false, false, false if we drop the word "prime." For example, consider the fraction $12/4$: 6 divides the numerator 12 and not the denominator 4, but it certainly doesn't divide the quotient $12/4 = 3$.

11.3 Binomial Coefficients Again

In the first part of this book we saw many fascinating patterns in Pascal's triangle, ranging from the simple and basic observation that each number is the sum of the two above it, to more complicated relations. We're going to take up Pascal's triangle once more here, again with an eye toward spotting a pattern. But it'll be a different sort of pattern we're looking for: we're going to look at the *divisibility* of the binomial coefficients.

To start with, let's reproduce here a sort of slimmed-down Pascal's triangle: we'll omit the 1s that occur along the edges, and just give the middle terms.

2
3 3
4 6 4
5 10 10 5
6 15 20 15 6
7 21 35 35 21 7
8 28 56 70 56 28 8
9 36 84 126 126 84 36 9
10 45 120 210 252 210 120 45 10
11 55 165 330 462 462 330 165 55 11

Now, the pattern we're looking for here is particularly hard to spot, for two reasons. One, it's not the same kind of pattern we observed before: it doesn't have to do with combinations or differences of binomial coefficients, but with individual ones, and has to do with the divisibility of the binomial coefficients rather than their size. The other reason is that it doesn't occur all the time. But here's a hint: look at the rows that start with 2, 3, 5, 7 or 11.

The answer is that *the binomial coefficients in those rows are all divisible by the first number in the row*—to say it in more technical language, if p is a prime and k is any number between 1 and $p-1$, then the binomial coefficient $\binom{p}{k}$ is divisible by p. Note that this isn't true at all of the other rows: in the "4" row, for example, there's the binomial coefficient $\binom{4}{2} = 6$ which isn't divisible by 4; in the "6" row there are the binomial coefficients $\binom{6}{2} = 15$ and $\binom{6}{3} = 20$, which aren't divisible by 6, and so on.

Now, we may observe this as a pattern that holds true for the primes $p = 2, 3, 5, 7$ and 11. We might suspect on the basis of this evidence that it's true for any prime, but how would we go about convincing ourselves of this fact?

Well, to show that $\binom{p}{k}$ is divisible by p, we start with the formula

$$\binom{p}{k} = \frac{p(p-1)(p-2)\ldots(p-k+1)}{k(k-1)(k-2)\ldots 1}.$$

We note that p appears once in the numerator, and does not divide any of the factors in the denominator. Thus the numerator n is divisible by p, while the denominator b is not; and, as we saw in the last section, that means that p must divide the quotient; in other words, $\binom{p}{k}$ must be divisible by p for $k = 1, 2, \ldots, p - 1$, as we suspected.

By the way, almost *none* of this argument works if p is replaced by a nonprime number n: when we ask which binomial coefficients $\binom{n}{k}$ are divisible by n, we are lost. We still have the formula

$$\binom{n}{k} = \frac{n(n-1)(n-2)\ldots(n-k+1)}{k(k-1)(k-2)\ldots 1}$$

but it doesn't tell us much. For one thing, if n is not prime, it's possible for a product of numbers to be divisible by n even if none of the individual factors are; this means that even though the terms in the denominator are each smaller than n, their product may well be divisible by n.

What's more, even if the denominator were not divisible by n and the numerator were, it wouldn't follow that the quotient was divisible by n: as we observed before, 6 divides the numerator of the fraction 12/4 and not the denominator, but it doesn't divide the quotient. You can see this in the table on the last page: if you look at the $n = 4$, 6, 8, 9 or 10 row, you'll see entries that are not divisible by n.

In fact, it's very difficult to give a general rule for when a binomial coefficient $\binom{n}{k}$ is divisible by n, except when n is a prime.

11.4 The Fundamental Theorem of Arithmetic

The principal result of Section 11.2—that a prime divides a product only if it divides one of the factors—leads in turn to a basic fact about numbers.

We've seen that any number can be written as a product of primes: for example

$$35 = 5 \times 7$$

$$741 = 3 \times 13 \times 19$$

and

$$14{,}467 = 17 \times 23 \times 37.$$

We now want to ask the question: *can there be more than one way of writing a given number as a product of primes?* For example, can you write 35 as a product of primes other than 5 and 7? The answer in that case is pretty clear: it doesn't take long to check that no prime other than 5 and 7 divides 35. But what if we asked the same question for 14,467? What if we told you, for example, that 31 divided 14,467. Would you believe us?

No! You shouldn't if you believed us the first time, when we said that $14{,}467 = 17 \times 23 \times 37$. Because *if* 31 *divided* 14,467, *it would have to divide one of the factors* 17, 23 *or* 37 of 14,467, and it doesn't. In fact, if any prime number p divided 14,467, it would have to divide 17, 23 or 37, and since these numbers are all prime—that is, they're not divisible by any number besides 1 and themselves—p would have to be one of them. In other words, once we know that $14{,}467 = 17 \times 23 \times 37$, it follows that the *only* primes dividing 14,467 are 17, 23 and 37

The point is, this works in general: once we've expressed a number n as a product of primes, then *any* prime dividing n must be among the factors. So there can't be two essentially different ways of expressing a number as a product of primes: that is,

> Any number can be written as a product of primes in one and only one way.

This statement is called the **Fundamental Theorem of Arithmetic**, and while that might sound like a fairly overblown title for such a simple statement, it doesn't do more than simple justice to the absolutely basic role it plays in our understanding of numbers. We'll see in the next chapter some of the many consequences of this statement, but for now let's just observe that one basic fact about the Theorem: it gives us a second way of expressing numbers, and one that reveals much more about the number than the standard decimal one. Instead of writing 18, for example, we can write its prime factorization $2 \cdot 3^2$; instead of 84 we can write $2^2 \cdot 3 \cdot 7$; instead of 616, $2^3 \cdot 7 \cdot 11$ and so on.

When you come right down to it, in many ways it makes a lot more sense than the conventional way of representing numbers. Take for example the number 384, which has prime factorization $2^7 \cdot 3$. When you write "384," you convey the information that the number in question is

$$3 \cdot 10^2 + 8 \cdot 10 + 4.$$

In effect, you're giving a sort of "recipe" in additive terms for making 384: you're saying, "take three 100s, eight 10s and four 1s and add them all up." What we have now is another sort of recipe for 384, this time a multiplicative one: when we write

$$2^7 \cdot 3$$

we're saying, "take seven 2s and a 3, and multiply them all together." As you might expect (and as we'll see amply demonstrated in the following chapter), you can usually tell much more about the arithmetic properties of the number from this representation.

Of course, this way of expressing numbers is not uniformly superior. It's hard, for example, to compare the relative size of numbers written in this way: for example, is $2^7 \cdot 3$ more or less than $5 \cdot 7 \cdot 11$? Also, addition, which is a pretty simple task when we represent numbers decimally, is a mess in this new notation. But that's usually the way it is with a second language: some things are better expressed in the new language, some in the old.

We'll see how to systematically exploit some of the properties of this new language in the following chapter, but first we'd like to try and explain by analogy why it occupies such a central role.

11.5 It's a Matter of Chemistry

In elementary chemistry we're taught that all matter is made up of molecules. Molecules are in turn made up of atoms, which are classified into the elements: hydrogen, oxygen, carbon, sodium, chlorine and so on. Each element is denoted by its own symbol, consisting of one or two letters—the five elements above, for example, are denoted H, O, C, Na and Cl respectively.

The composition of a molecule is specified by saying how many atoms of each type it contains: a molecule of water, for example, consists of two hydrogen atoms and one oxygen atom, so that we write it as H_2O. A molecule of carbon dioxide is made up of one carbon atom and two oxygen: CO_2.

When we combine two types of molecules, they may break up into their constituent atoms and regroup into other compound molecules. For example, a molecule of sodium bicarbonate (baking soda) is made up of a sodium atom, a hydrogen atom, a carbon atom and three oxygen atoms: $NaHCO_3$ altogether. Hydrochloric acid is simpler: a molecule consists simply of one atom each of hydrogen and chlorine, and so is denoted HCl. Now, if we mix sodium bicarbonate and hydrochloric acid, the chlorine and sodium atoms will combine to form sodium chloride (NaCl, or table salt); the carbon atom will merge with two of the oxygen atoms to form carbon dioxide, and the two hydrogen atoms will combine with the remaining oxygen atom to form water. In chemical notation,

$$\mathrm{NaHCO_3} + \mathrm{HCl} \longrightarrow \mathrm{NaCl} + \mathrm{CO_2} + \mathrm{H_2O}.$$

Notice one thing: a chemical reaction—like this one, for example—may produce a compound like water, or carbon dioxide, even if none of the initial ingredients contained water or carbon dioxide. But an *element* can only appear in the output of a reaction if it is contained in one of the ingredients.

The analogy to what we've been doing in this chapter is pretty clear. If numbers in general are like compounds, then primes are the elements of which they're composed. If we combine (multiply) two composite numbers, we can rearrange the factors to express the result as a product of other numbers: $14 \times 15 = 2 \times 7 \times 3 \times 5 = 10 \times 21$. For a direct analogy with the reaction above, associate to the elements H, O, C, Na and Cl the (randomly chosen) primes 2, 3, 5, 7 and 11. Then the compounds $NaHCO_3$ and HCl correspond to the composite numbers $7 \times 2 \times 5 \times 3^3 = 1{,}890$ and $2 \times 11 = 22$; the compounds NaCl, CO_2 and H_2O would correspond to $7 \times 11 = 77$, $5 \times 3^2 = 45$ and $2^2 \times 3 = 12$; and the chemical reaction above would correspond to the equality

$$1{,}890 \times 22 = 41{,}580 = 77 \times 45 \times 12.$$

In these terms, the statement that "an element can only appear in the output of a reaction if it's contained in one of the ingredients" corresponds to our fact that a prime divides a product only if it divides one of the factors.

The point of this analogy is not that it'll help you do either arithmetic or chemistry better. It's to help you understand the absolutely basic role that primes—and the Fundamental Theorem of Arithmetic—play in the study of numbers. Primes are the building blocks of all numbers in exactly the way that the elements are the building blocks of all matter. Likewise, the most crucial fact you can know about any number is its prime factorization, in much the same way as the most important information about any compound is what elements make it up. (Historically, it's only fair to point out, the concept of "elements" appeared first in mathematics, as the primes. The atoms of physics and the elements of chemistry came later and were modeled on this mathematical theory.)

In line with this, ask yourself: what is the most basic, the very first thing that anyone sees in a chemistry course? What information is on the wall of every chemistry classroom at every level, from elementary school to college? The answer, of course, is the *periodic table*, which lists the elements and suggests some patterns in their behavior. Now imagine, if you can, a chemist without knowledge of the periodic table—basically, it's inconceivable.

At this point, you should have some idea of the enormous importance that primes have for all mathematicians, and especially those who deal with numbers. Because when you come right down to it, we *don't* know the "periodic table" of our subject: where scientists have compiled a more or less complete list of all the elements found in nature (in our world, at least), we know that we can never have such a complete list in mathematics. And as much as we study the primes—the patterns in their distribution, their different behaviors—what we know about them remains minuscule compared to what we don't know.

Chapter 12

Consequences

As we said in our initial discussion of the Fundamental Theorem of Arithmetic, what it gives us is an alternative way of expressing numbers. Rather than expressing a number as a sum of multiples of powers of 10, as we usually do, we express it as a product of powers of primes; in this language, for example, 756 is $2^2 \cdot 3^3 \cdot 7$.

In this chapter, we're going to do what we want to do whenever we learn a new language: we're going to learn the basic rules of grammar; we'll see how to reexpress some familiar concepts in terms of our new notation and we'll also learn some new things that are particularly easy to express in the new terms. Finally, in the last section, we'll discuss briefly the difficulties of translation, a topic we'll come back to later.

12.1 Preliminaries

Before we begin to work in earnest with the representation of numbers as products of primes, there are a couple of basic rules for combining powers of a number that we need to review.

The first is pretty obvious. If we multiply the ath power of a number x by the bth power, the result is simply the product of x with itself $a + b$ times—that is, the $(a + b)^{\text{th}}$ power of x:

$$\underbrace{(x \times x \times \cdots \times x)}_{a} \times \underbrace{(x \times \cdots \times x)}_{b} = \underbrace{x \times x \times x \times \cdots \times x \times x}_{a+b}$$

or in other words

$$x^a \cdot x^b = x^{a+b}.$$

For the second, suppose we take the ath power of a number x, and then raise that number to the bth power. We can write this out as the product of a string of x's

$$\underbrace{\underbrace{(x \times \cdots \times x)}_{a} \times \underbrace{(x \times \cdots \times x)}_{a} \times \ldots \underbrace{(x \times \cdots \times x)}_{a}}_{b} = \underbrace{(x \times x \times x \times \cdots \times x \times x)}_{ab}$$

or in other words,

$$(x^a)^b = x^{ab}.$$

Now, these two formulas are pretty straightforward, but they bear repeating. In particular, it's important to think of these as *two-way formulas*: when we see an expression like $(x^4)^6$ we know we can multiply this out to get x^{24}; but it'll also be very useful, when we see an expression like x^{24}, to remember that this can also be rewritten as $(x^4)^6$, or $(x^6)^4$, or $(x^8)^3$, and so on.

12.2 Multiplication and Division

One nice aspect of the representation of numbers by their prime factorizations is that the operations of multiplication and division become very simple. For example, it may not be obvious what the product of, say, 33,880 and 31,350 is. But if we write them as

$$33{,}880 = 2^3 \cdot 5 \cdot 7 \cdot 11^2$$

and

$$31{,}350 = 2 \cdot 3 \cdot 5^2 \cdot 11 \cdot 19$$

then the product is obtained simply by adding corresponding exponents in these expressions:

$$\begin{aligned} 33{,}880 \times 31{,}350 &= (2^3 \cdot 5 \cdot 7 \cdot 11^2) \cdot (2 \cdot 3 \cdot 5^2 \cdot 11 \cdot 19) \\ &= (2^3 \cdot 2) \cdot 3 \cdot (5 \cdot 5^2) \cdot 7 \cdot (11^2 \cdot 11) \cdot 19 \\ &= 2^4 \cdot 3 \cdot 5^3 \cdot 7 \cdot 11^3 \cdot 19. \end{aligned}$$

To give the general rule, we need to introduce some notation for expressing a number as a product of primes. Unfortunately, this is going to involve some indices with subscripts, which we'd prefer to avoid. But the actual math isn't that complicated; if you just bear a concrete example in mind you'll be fine. Anyway, suppose we have any positive whole number n. We can write n as a product of the form

$$n = 2^{a_2} \cdot 3^{a_3} \cdot 5^{a_5} \cdot 7^{a_7} \cdot \ldots,$$

for some collections of exponents $a_2, a_3, a_5, a_7, \ldots,$ which will all be whole numbers, either zero or positive. For example, if n were the number $33{,}880 = 2^3 \cdot 5 \cdot 7 \cdot 11^2$ above, a_2 would be 3; a_3 would be 0; a_5 and a_7 would each be 1; a_{11} would be 2 and all the others—a_{13}, a_{17}, a_{19} and so on—would be 0. Similarly, if we wrote the number $31{,}350 = 2 \cdot 3 \cdot 5^2 \cdot 11 \cdot 19$ above in the form

$$m = 2^{b_2} \cdot 3^{b_3} \cdot 5^{b_5} \cdot 7^{b_7} \cdot \ldots$$

then the exponents would be

$$b_2 = 1$$
$$b_3 = 1$$
$$b_5 = 2$$
$$b_{11} = 1$$
$$b_{19} = 1$$

and all the rest would be zero:

$$b_7 = b_{13} = b_{17} = b_{23} = \cdots = 0.$$

Now, let's go back to giving formulas for operations on numbers in factored form. Suppose we have two numbers n and m, which are written as

$$n = 2^{a_2} \cdot 3^{a_3} \cdot 5^{a_5} \cdot 7^{a_7} \cdot \ldots,$$

and

$$m = 2^{b_2} \cdot 3^{b_3} \cdot 5^{b_5} \cdot 7^{b_7} \cdot \ldots.$$

To start with, the product is obtained by adding exponents:

$$n \cdot m = 2^{a_2+b_2} \cdot 3^{a_3+b_3} \cdot 5^{a_5+b_5} \cdot 7^{a_7+b_7} \cdot \ldots.$$

Similarly, if we raise a number to a power, the second of our basic rules for exponents tells us that we simply multiply the exponents by the power: for example, if

$$n = 2^{a_2} \cdot 3^{a_3} \cdot 5^{a_5} \cdot 7^{a_7} \cdot \ldots$$

as before, then its square is

$$n^2 = 2^{2a_2} \cdot 3^{2a_3} \cdot 5^{2a_5} \cdot 7^{2a_7} \cdot \ldots;$$

its cube is

$$n^3 = 2^{3a_2} \cdot 3^{3a_3} \cdot 5^{3a_5} \cdot 7^{3a_7} \cdot \ldots,$$

and so on.

The rule for multiplication leads in turn to other rules. For example, what does it mean to say that a number m divides a number n? It means there's a whole number you can multiply m by to get n. But by the rule we just saw, if that's the case then each exponent in the prime factorization of n must be at least as large as the corresponding exponent for m. In other words, we have the rule: if

$$n = 2^{a_2} \cdot 3^{a_3} \cdot 5^{a_5} \cdot 7^{a_7} \cdot \ldots,$$

and

$$m = 2^{b_2} \cdot 3^{b_3} \cdot 5^{b_5} \cdot 7^{b_7} \cdot \ldots.$$

are any two numbers, then m divides n precisely when $a_2 \geq b_2$, $a_3 \geq b_3$, $a_5 \geq b_5$ and so on; and in this case

$$\frac{n}{m} = 2^{a_2-b_2} \cdot 3^{a_3-b_3} \cdot 5^{a_5-b_5} \cdot 7^{a_7-b_7} \cdot \ldots.$$

Exercise 12.2.1 For each of the following pairs of numbers, say whether m divides n, and if it does find the quotient n/m:

1. $m = 2^2 \cdot 3 \cdot 7$ and $n = 2^3 \cdot 3 \cdot 7^2 \cdot 13$
2. $m = 2^2 \cdot 3 \cdot 11$ and $n = 2^4 \cdot 11 \cdot 19$
3. $m = 2^3 \cdot 3 \cdot 7$ and $n = 2^2 \cdot 3 \cdot 7^2 \cdot 13$
4. $m = 2^2 \cdot 3 \cdot 7 \cdot 11 \cdot 13$ and $n = 2^3 \cdot 3 \cdot 7^2 \cdot 11^2 \cdot 13$

12.3 How Many Numbers Divide 756?

One thing we can do very readily with our new way of representing numbers is to count the number of divisors a number has. We'll show you how to do this in an example.

EXAMPLE 12.3.1 How many numbers (including 1 and 756) divide 756?

SOLUTION First of all, we find the prime factorization of 756. This is not hard: to begin with, it's clearly even, so we can write

$$756 = 2 \times 378.$$

378 is again even, so we divide by 2 again:

$$378 = 2 \times 189.$$

The next prime to look at is 3, and indeed 189 is divisible by 3: we have

$$189 = 3 \times 63$$

and again

$$63 = 3 \times 21$$

and finally

$$21 = 3 \times 7.$$

So we're done: we have

$$756 = 2^2 \cdot 3^3 \cdot 7.$$

Now we're ready to ask what numbers n divide 756. We know the answer: if a number n divides 756, its prime factorization can involve only the primes 2, 3 and 7, and can have exponents at most 2, 3 and 1 respectively—that is, any number n dividing 756 must look like

$$n = 2^a \cdot 3^b \cdot 7^c$$

with

$$a \leq 2, \qquad b \leq 3 \quad \text{and} \quad c \leq 1.$$

What this means is that in order to specify a number n dividing 756, we have to choose first a number a between 0 and 2; then a number b between 0 and 3, and finally a number c between 0 and 1. By the multiplication principle we introduced way back in Chapter 2, then, the total number of such choices is

$$3 \times 4 \times 2 = 24.$$

We conclude, then, *that there are exactly 24 numbers dividing 756.* ■

Now it's your turn:

Exercise 12.3.2 Find the number of divisors of each of the following numbers:

1. 104
2. 210
3. 384

12.4 gcd's and lcm's

We can use the criterion for divisibility we found in Section 12.2 to describe the least common multiple and greatest common divisor of two numbers. Let's start with an example: suppose we're asked to find the greatest common divisor of the number

$$756 = 2^2 \cdot 3^3 \cdot 7$$

and another number whose factorization we know, say

$$264 = 2^3 \cdot 3 \cdot 11.$$

Now, in steps: we've just seen

- If a number divides 756, its prime factorization involves only 2, 3 and 7, and the exponents of 2, 3 and 7 are at most 2, 3 and 1 respectively.
- If a number divides 264, its prime factorization involves only 2, 3 and 11, and the exponents of 2, 3 and 11 are at most 3, 1 and 1 respectively. It follows that
- If a number divides *both* 756 and 264, its prime factorization involves only 2 and 3, and the exponents of 2 and 3 are at most 2 and 1 respectively. Therefore:
- The greatest common divisor of 756 and 264 is $2^2 \times 3 = 12$.

The same sort of logic tells us the least common multiple as well: following the same steps, we know that

- If a number is divisible by 756, its prime factorization must involve 2, 3 and 7 with exponents at least 2, 3 and 1 respectively.
- If a number is divisible by 264, its prime factorization must involve 2, 3 and 11 with exponents at least 3, 1 and 1 respectively. It follows that
- If a number is divisible by *both* 756 and 264, its prime factorization must involve 2, 3, 7 and 11, with exponents at least 3, 3, 1 and 1 respectively. Therefore:
- The least common multiple of 756 and 264 is $2^3 \times 3^3 \times 7 \times 11 = 16{,}632$.

It's pretty clear how this goes in general, sufficiently so that we're willing to risk a little notation. Suppose we're given two numbers n and m, with prime factorizations

$$n = 2^{a_2} \cdot 3^{a_3} \cdot 5^{a_5} \cdot 7^{a_7} \cdot \ldots$$

and

$$m = 2^{b_2} \cdot 3^{b_3} \cdot 5^{b_5} \cdot 7^{b_7} \cdot \ldots.$$

Then the greatest common divisor of n and m will be the number

$$\gcd(n, m) = 2^{c_2} \cdot 3^{c_3} \cdot 5^{c_5} \cdot \ldots,$$

where c_2 is the smaller of the two numbers a_2 and b_2, c_3 is the smaller of the two numbers a_3 and b_3, and so on. Similarly, the least common multiple of the two will be

$$\text{lcm}(n, m) = 2^{d_2} \cdot 3^{d_3} \cdot 5^{d_5} \cdot \ldots,$$

where d_2 is the greater of the two numbers a_2 and b_2, d_3 is the greater of the two numbers a_3 and b_3, and so on.

So we see that if we know the prime factorizations of two numbers we can find their least common multiple and greatest common divisor pretty easily. Now, this is not usually a practical way of computing these quantities, for reasons we'll discuss in Section 12.7 below. But it does tell us an important fact about the least common multiple and greatest common divisor of two numbers.

This is based on a seemingly trivial observation. To see it, suppose a and b are any two numbers. Now let c be the smaller of the two, and d be the larger of the two. Then of course the two numbers c and d are just the two numbers a and b, possibly in reverse order; in any event,

$$c + d = a + b.$$

Now apply this to the expressions we just worked out for the least common multiple and greatest common divisor of two numbers n and m as above: it says that

$$c_2 + d_2 = a_2 + b_2$$
$$c_3 + d_3 = a_3 + b_3$$
$$c_5 + d_5 = a_5 + b_5$$

and so on. Which means that when you multiply the least common multiple and greatest common divisor, you get

$$\begin{aligned}\text{lcm}(n, m) \cdot \gcd(n, m) &= 2^{c_2+d_2} \cdot 3^{c_3+d_3} \cdot 5^{c_5+d_5} \cdot 7^{c_7+d_7} \cdot \ldots \\ &= 2^{a_2+b_2} \cdot 3^{a_3+b_3} \cdot 5^{a_5+b_5} \cdot 7^{a_7+b_7} \cdot \ldots \\ &= n \cdot m.\end{aligned}$$

In other words, we've now established for any pair of numbers n and m the rule we discovered empirically back in Section 8.3:

$$\text{lcm}(n, m) \cdot \gcd(n, m) = m \cdot n.$$

Or, as we expressed it back then, *the least common multiple of two numbers is their product divided by their greatest common divisor.*

Exercise 12.4.1 For each of the following pairs of numbers, find their greatest common divisor and their least common multiple, and verify that the product of these is equal to the product of the original pair. For the first, verify your calculation of the greatest common divisor by applying the Euclidean algorithm.

1. $384 = 2^7 \cdot 3$ and $864 = 2^5 \cdot 3^3$
2. $56 = 2^3 \cdot 7$ and $325 = 5^2 \cdot 13$
3. $104{,}500 = 2^2 \cdot 5^3 \cdot 11 \cdot 19$ and $83{,}490 = 2 \cdot 3 \cdot 5 \cdot 11^2 \cdot 23$

Exercise 12.4.2

1. Find the prime factorization of 36,764 and 30,240.
2. Determine the greatest common divisor of 36,764 and 30,240 by using the Euclidean algorithm.
3. Determine the greatest common divisor of 36,764 and 30,240 by using the prime factorization you found in the first part.

Exercise 12.4.3 How many numbers divide both 130,000 and 13,600? (Hint: $130{,}000 = 2^4 5^4 13$ and $13{,}600 = 2^5 5^2 17$)

Exercise 12.4.4 Consider the numbers $m = 2^8 3^5 7^3 17^4$ and $n = 2^5 5^4 7^2 11^3$.

1. How many positive whole numbers divide m?
2. How many numbers divide both m and n?

12.5 Factorials Again

In Section 11.3 we saw in particular that a binomial coefficient of the form $\binom{p}{k}$, with p a prime number and k any number between 1 and $p-1$, was divisible by p. In fact, we can use unique factorization to answer questions about the divisibility of products and quotients of factorials much more generally. We'll do an example, and then let you try.

EXAMPLE 12.5.1 Consider the binomial coefficient

$$c = \binom{21}{10}.$$

Does 13 divide c? Does 5 divide c? How about 3, and 15?

SOLUTION Let's start with the easiest of these three questions, whether 13 divides c. To begin, we write the standard expression for the binomial coefficient c:

$$\binom{21}{10} = \frac{21!}{10!11!}.$$

Now, the numerator 21! of this expression is the product of all the numbers from 1 up to 21, which includes 13; so 13 clearly divides it. (In fact, since 13 appears as a factor of 21!, but doesn't divide any other factor, we see that in the prime factorization of the numerator 13 will appear exactly once.)

On the other hand, the denominator 10!11! of the expression is a product of numbers none of which is divisible by 13; and because 13 is prime, if it doesn't divide any of the factors of 10!11! then it can't divide 10!11! itself. Thus we see that 13 divides the numerator, but not the denominator; and—again using the fact that 13 is prime, and the principle expressed at the end of Section 11.2—it follows that 13 must divide the quotient as well.

We can similarly say how many times the factor 5 appears in the prime factorization of the numerator and denominator, and so determine if the quotient is divisible by 5 or not. Specifically, we see that four of the factors in the numerator—5, 10, 15 and 20—are divisible by 5, and each is divisible by 5 exactly once; so 5 will appear to the 4th power in the prime factorization of the numerator. At the same time, we see that 5 divides four of the factors in the denominator once, so 5 will appear to the 4th power in the prime factorization of the denominator. It follows that 5 will not appear at all in the prime factorization of the quotient; that is, 5 doesn't divide $\binom{21}{10}$.

The case of 3 is more complicated, but it goes the same way. First, we count how many 3s there are in the prime factorization of 21!. We see that 3 divides the seven factors

$$3, \quad 6, \quad 9, \quad 12, \quad 15, \quad 18, \quad \text{and} \quad 21$$

and moreover it divides two of them—9 and 18—twice. So 3 will appear with exponent $1 + 1 + 2 + 1 + 1 + 2 + 1 = 9$ in the prime factorization of 21!. As for the denominator, 3 divides six of the factors—the 3, 6 and 9 in each factorial—and divides two of them twice; so 3 will appear with exponent 8 in the prime factorization of 10!11!, and it follows that 3 will appear once in the prime factorization of $\binom{21}{10}$; in particular, it will divide $\binom{21}{10}$.

Finally, what about 15? Well, we see that 15 appears in the numerator 21!, but doesn't appear at all in the denominator 11!10!, so it must divide the quotient, right?

No! *That only works with primes!* That's the whole *point* of what it means to be a prime, in fact. We're sorry to go typographically apoplectic on you, but it's simply not true that if 15 doesn't divide any factor of 11!10! then it doesn't divide 11!10!; and this is a good example.

How do we decide whether 15 divides $\binom{21}{10}$? Well, if you think about it, a number will be divisible by 15 exactly when it's divisible by both 3 and 5, and we just saw a moment ago that 5 doesn't divide $\binom{21}{10}$, so 15 can't either. In general, if we want to decide whether a composite number n divides a product or quotient, we have to see if each of the prime factors of n divides it, and how many times. ■

Exercise 12.5.2 Consider the following fraction:

$$c = \frac{31!}{12! \cdot 17!}.$$

1. Show that c is a whole number.
2. Does 7 divide c? Explain why or why not.
3. Does 13 divide c? Again, explain why or why not.
4. Does 17 divide c? Again, explain why or why not.
5. Does 91 divide c? Again, explain why or why not.

Exercise 12.5.3 Which of the following numbers is divisible by 91?

$$\frac{44!}{21!21!} \qquad \frac{43!}{14!26!} \qquad \frac{43!}{20!21!} \qquad \frac{48!}{19!27!}$$

12.6 $\sqrt{2}$ Is Not a Fraction

As a final application of the ideas we've developed in this part of the book, we're going to answer a question posed (and answered) by both the Babylonians and the ancient Greeks: is there a fraction whose square is 2? In other words, can we find a pair of whole numbers n and m such that

$$\left(\frac{n}{m}\right)^2 = 2?$$

The answer, as we'll see in a moment, is no.

Now, there are plenty of fractions whose squares are close to 2: $\frac{7}{5}$ has square equal to 1.96, which isn't bad. For closer work we might try $\frac{17}{12}$, whose square is $\frac{289}{144}$, which is within $\frac{1}{100}$ of 2; or for really fine work we could use $\frac{239}{169}$, whose square, $\frac{57,121}{28,561}$, is within $\frac{4}{100,000}$ of 2. These fractional approximations to $\sqrt{2}$ are sufficient for almost all real world calculations. But the fact remains that there is no fraction whose square is *exactly* 2.

How can we be certain of that, especially given that there are fractions whose squares are so close? The key is to represent the numbers n and m we're looking for in terms of their prime factorization. Put it this way: suppose m and n are any two whole numbers. We want to know if it's possible that

$$\left(\frac{n}{m}\right)^2 = 2,$$

or, equivalently, if it's possible that

$$n^2 = 2 \cdot m^2.$$

Now, let's write the prime factorizations of m and n as

$$n = 2^{a_2} \cdot 3^{a_3} \cdot 5^{a_5} \cdot \ldots$$

and

$$m = 2^{b_2} \cdot 3^{b_3} \cdot 5^{b_5} \cdot \ldots.$$

We can then get the prime factorization of n^2 just by doubling the exponents in the prime factorization of n:

$$n^2 = 2^{2a_2} \cdot 3^{2a_3} \cdot 5^{2a_5} \cdot \ldots.$$

Likewise, we can write the square of m as

$$m^2 = 2^{2b_2} \cdot 3^{2b_3} \cdot 5^{2b_5} \cdot \ldots;$$

and then we can multiply by 2 to arrive at the expression

$$2m^2 = 2^{2b_2+1} \cdot 3^{2b_3} \cdot 5^{2b_5} \cdot \ldots.$$

Now, compare the two expressions for n^2 and $2m^2$. We see that *they can never be equal, because the exponent of 2 in the first is even, while in the second it's odd*. Thus there is no pair of whole numbers n and m with $n^2 = 2m^2$, and correspondingly no fraction whose square is 2.

Exercise 12.6.1 Go through a similar argument to convince yourself that

1. There is no fraction whose square is 3.
2. There is no fraction whose cube is 2.

In fact, the same argument can be used to show in general that a whole number can be the square of a fraction only when all its exponents are even—that is, when it's already the square of a whole number. To put it differently, a number that is not the square of a whole number can't be the square of a fraction either; in particular, *no prime number has a square root that is a fraction.*

Exercise 12.6.2 Which of the following fractions is the square of a fraction?

$$\frac{2^3 \cdot 7^4}{5^2 \cdot 11^6} \qquad \frac{2^6 \cdot 7^4}{5^2 \cdot 11^5} \qquad \frac{2^2 \cdot 7^7}{5^5 \cdot 11^{11}} \qquad \frac{2^8 \cdot 7^4}{5^2 \cdot 11^6}$$

12.7 The Fine Print

Here, unfortunately, is where we hit the "some restrictions may apply" part. To the extent that we've convinced you at this point that representing numbers by their prime factorization is at least occasionally useful, a natural question arises: how hard is it to go back and forth between the two expressions of a number?

In one direction, certainly, there's no problem: if we're given the prime factorization of a number, we can multiply it out easily enough; and even if the number in question is huge—say on the order of 100 digits—it wouldn't take more than a split second for even a desktop computer to carry out the calculation.

The other direction is another story entirely. If we asked you to find the factorization of a number like, for instance, 323, the procedure would be clear: check if it's divisible by 2 (it's not); check whether it's divisible by 3 (ditto); then check divisibility by 5, 7 and so on; each time, if the number we start with is divisible we divide and then proceed with the quotient. We would just have to check divisibility by primes less than $\sqrt{323}$, that is, up to 17. Even for those of us whose calculator fingers are not especially nimble, it wouldn't take more than a minute or so to find in this way that $323 = 17 \times 19$.

But how long would it take to check a 100-digit number n in this way? We'd have to check divisibility by numbers up to $\sqrt{n}$, which in this case is roughly 10^{50}. And while it's true we only have to check divisibility by primes, this doesn't gladden our hearts all that much, for two reasons. One, we don't have a list of all the primes up to 10^{50}. Two, when we discuss the distribution of primes at greater length, we will see that the number of primes less than 10^{50} is on the order of 10^{48}, which is still pretty large.

How long would it take us to perform 10^{48} divisions? Well, let's assume we have the fastest computers available today. Such computers can carry out roughly a trillion calculations each second. Of course, it takes quite a few calculations to divide a 100-digit number, but let's be generous and suppose that our computer can actually perform a trillion divisions a second. Since a trillion is 10^{12}, this means the process of factoring our randomly chosen number n of size 10^{100} could take

$$\frac{10^{48}}{10^{12}} = 10^{36} \text{ seconds.}$$

This is the same as

$$\frac{10^{36}}{86{,}400} \sim 10^{31} \text{ days,}$$

which is to say

$$\frac{10^{31}}{365} \sim 3 \times 10^{28} \text{ years}$$

or in other words

$$\frac{3 \times 10^{28}}{100} \sim 3 \times 10^{26} \text{ centuries.}$$

This is several orders of magnitude longer than the lifetime of the universe!

There are other methods of factoring a number n, which are considerably faster than trial division by all prime numbers up to the square root of n. Using these methods, and thousands of linked computers, mathematicians have recently been able to factor numbers of up to 150 digits in a reasonable amount of time (less than a month). But it seems at the present moment that numbers of 500 digits are completely unreachable: no algorithm is known for factoring one that doesn't potentially involve waiting around until the sun is a black, lifeless lump of coal, and no realistic advance in the speed of computers—even, say, a billionfold improvement—will bring it into range.

By the way, the word "seems" in the last paragraph is a critical one. Right now, not only don't we have an algorithm that will factor a 500-digit number in a reasonable time, *we don't even know whether such an algorithm is theoretically possible*. Someone could have a new idea and come up with a method tomorrow. Or someone might prove that no such algorithm is possible; that is, that factorization is an intrinsically "hard" problem. (We would bet that this is in fact the case.) The question of whether or not such an algorithm exists is one of the major outstanding problems in mathematics. Given the fact that—as we'll see in the last part of the book—many of the cryptography systems currently in use rely for their security on the impossibility of factoring large numbers, it's an important question for all of us.

To sum up: the fact that any number has a unique prime factorization is vital to our understanding of what numbers are. It's also a tremendously useful tool for proving formulas about numbers, and for understanding their behavior in general. But it's not always a practical tool for dealing with specific large numbers.

JASON, WHAT PATTERN DID I TELL YOU TO RUN?
YOU SAID TO GO 10 YARDS OUT, THEN 10 YARDS TO THE RIGHT.
AND WHAT PATTERN DID YOU RUN?
I WENT 10√2 YARDS AT A 45-DEGREE ANGLE.
AND ARE THEY THE SAME THING??
IF YOU ADD THE VECTORS, SURE.
THIS IS FOOTBALL, NOT MATH CLASS!
WHY DO YOU THINK I ROUNDED TO TWO DECIMAL PLACES?
www.foxtrot.com
© 1999 Bill Amend/Dist. by Universal Press Syndicate
11-14
AMEND

Chapter 13

Relatively Prime

In the last chapter of this part of the book, we're going to take up a notion related to primality and prime factorization: when two numbers are *relatively prime*. We'll say first what this means, and then go on to pose and solve a counting problem that turns out to be absolutely fundamental: what proportion of all numbers are relatively prime to a given number?

13.1 What Does It Mean to Be Relatively Prime?

We say that two numbers m and n are *relatively prime* if their greatest common divisor is 1—that is, if no whole number bigger than 1 divides both m and n. To hammer the point home, here are a number of equivalent ways of saying it. To start with the definition, m and n are relatively prime if

- $\gcd(m, n) = 1$

which means precisely that

- the only positive whole number dividing both m and n is 1.

Now, since any number greater than 1 that divided both m and n would in turn be divisible by at least one prime, to say that m and n are relatively prime is equivalent to saying that

- no prime number divides both m and n,

or, in other words,

- the prime factorizations of m and n have no primes in common.

By the way, here's an interesting consequence of this way of putting it. As we've seen, the prime factorization of powers of a number n involve the same primes as the factorization of n itself, just to higher powers. So if the prime factorizations of two numbers m and n have no primes in common, the same is true of any powers of m and n—in other words, if m and n are relatively prime, then so are any powers of m and n, and conversely.

Next, recalling our discussion of divisibility earlier in this part of the book, we have a couple more equivalent formulations. First, we can say that m and n are relatively prime if

- $\operatorname{lcm}(m, n) = mn$,

which in turn means precisely that

- any number divisible by both m and n is a multiple of the product mn.

Lastly, from what we saw in the discussion of combinations, we can say that m and n are relatively prime if

- the number 1 is a combination of m and n;

or, equivalently, if

- every whole number is a combination of m and n.

Let's take a moment out here to notice something that may seem kind of mysterious. We observed earlier that if two numbers are relatively prime then so are any of their powers. If we combine that observation with these last characterizations of what it means to be relatively prime, we arrive at a syllogism: for any pair of numbers m and n, if we can express 1 as a combination of m and n, then m and n must be relatively prime; so any powers of m and n are relatively prime; so it must be possible to express 1 as a combination of any powers of m and n. In other words, for any numbers m, n, a and b,

> If we can express 1 as a combination of m and n, then we can express 1 as a combination of m^a and n^b.

Now, stop and think about that statement for a minute. The first thing to see is that *it's not at all obvious that this is true*. For example, the statement tells us that because we can express 1 as a combination of 2 and 3, it must be possible to express 1 as a combination of any powers of 2 and 3—for instance, it must be possible to express 1 as a combination of $2^8 = 256$ and $3^6 = 729$. You have to admit, it's not obvious. What's more, knowing that it's possible to express 1 as a combination of 256 and 729 doesn't give us a leg up on doing it—if we wanted to actually find a solution of the equation

$$x \cdot 256 + y \cdot 729 = 1$$

we'd have to go through the whole process described in Section 9.4, just as we would for any pair of numbers in place of 256 and 729.

This is a good illustration of the hidden power of the Euclidean algorithm and prime factorization: combining seemingly simple facts, we arrive at a conclusion like the boxed statement above that is far from clear. It's also a good example of a situation that arises often in math, where logic may tell us that a solution to a certain problem exists without giving us any special knowledge of how to go about solving it.

Had enough? Do the following exercise (it's easy) and then we'll get down to a real question.

Exercise 13.1.1 Say whether each of the following pairs of numbers are relatively prime.

1. 66 and 70
2. 96 and 105
3. 234 and 399
4. 7 and 43,784
5. 64 and 32,965

13.2 The Euler ϕ-function

The problem we're going to solve is simple enough to state: we're going to pick a number n and ask, "How many of the numbers between 0 and $n-1$ are relatively prime to n?" The answer is usually denoted $\phi(n)$. (This is called the *Euler ϕ-function*, after the 18th century mathematician Leonhard Euler, who first discussed it in connection with a result we'll encounter in the next part of the book.) Here are some examples:

$n = 6$: Of the numbers 0, 1, 2, 3, 4 and 5, only 1 and 5 are relatively prime to 6; so $\phi(6) = 2$.

$n = 8$: Since the only prime dividing 8 is 2, the numbers relatively prime to 8 are just the odd numbers. In particular, of the eight numbers $0, 1, 2, \ldots, 7$, the ones relatively prime to 8 are 1, 3, 5 and 7; so $\phi(8) = 4$.

$n = 13$: Since 13 is prime, any number not divisible by 13 is relatively prime to it. In particular, the numbers $1, 2, 3, \ldots, 12$ are all relatively prime to 13, so $\phi(13) = 12$. In general, if p is any prime number, all the numbers between 1 and $p-1$ are relatively prime to it, so $\phi(p) = p - 1$.

$n = 60$: We know that $30 = 2^2 \cdot 3 \cdot 5$, so the numbers relatively prime to 60 are simply those not divisible by 2, 3 or 5. Here's a list of those between 1 and 59:

$$1, 7, 11, 13, 17, 19, 23, 29, 31, 37, 41, 43, 47, 49, 53, 59$$

so $\phi(60) = 16$.

Now you do some:

Exercise 13.2.1 Find the following:

1. $\phi(18)$
2. $\phi(21)$
3. $\phi(27)$
4. $\phi(37)$

How should we go about finding $\phi(n)$ in general? For small values of n we can grind it out: check each number between 1 and $n-1$ in turn to see if it's relatively prime to n, make a list of those that are and count the list. But as you can imagine that gets old pretty fast. We'll try instead to be more systematic: we're going to use what we might call a *modified sieve*.

Remember that in the sieve of Eratosthenes, in order to find the prime numbers we wrote out a list of all numbers and crossed off in turn the numbers divisible by 2, 3, 5, 7 and so on. To find the numbers relatively prime to a number n, we're going to do the same thing, but this time we're only going to cross off the numbers divisible by one of the primes dividing n. For example, to find the numbers between 0 and 14 relatively prime to 15, we start with the list

0 1 2 3 4 5 6 7 8 9 10 11 12 13 14.

Now, since $15 = 3 \cdot 5$, to say that a number is relatively prime to 15 just means that it's not divisible by either 3 or 5; so we want to cross off the numbers that are multiples of 3 or 5. We'll start with multiples of 3:

~~0~~ 1 2 ~~3~~ 4 5 ~~6~~ 7 8 ~~9~~ 10 11 ~~12~~ 13 14

Next we cross off the numbers divisible by 5:

~~0~~ 1 2 ~~3~~ 4 ~~5~~ ~~6~~ 7 8 ~~9~~ ~~10~~ 11 ~~12~~ 13 14

Of course, at this point we can simply count the numbers left to see that $\phi(15) = 8$. But we want a recipe for $\phi(n)$ that's going to work in general, without going through this process in full each time. So we should go back over this and ask: *at each stage of this process, how many numbers did we cross off?*

Well, that's easy enough to answer at the first stage: we start with 15 numbers and cross off every third number, so of course the number crossed off is $1/3$ of 15, or five; the number we have left after the first stage is correspondingly $2/3$ of 15, or 10.

What about at the second stage? Well, of the 15 numbers we started with, exactly $1/5$ of them, or three, are crossed off at the second stage; $4/5$ of them, or 12, are spared. But, you'll say, (correctly) that can't be the answer, since we have double crossings: what we really need to know is not how many of the original 15 are crossed off at the second stage, but how many of those that are left after the first stage are axed at the second stage.

Now we see something really interesting. Of the 10 numbers left after the first pass, exactly two are crossed off in the second: in other words, the *fraction* $1/5$ of the numbers left that are thrown out is exactly the same as the fraction of the original 15 that are hit. The number left at the end is correspondingly $4/5$ of the number left after the first go round, so we see that

$$\phi(15) = 15 \cdot \frac{2}{3} \cdot \frac{4}{5} = 8.$$

Exercise 13.2.2 Try doing the same process in the other order, that is, crossing off first the numbers between 0 and 14 divisible by 5, and then crossing off those divisible by 3. What fraction of the numbers are crossed off at the first step? What fraction *of the numbers left* are crossed off at the second?

Now let's see if this works in another case, say $n = 18$. To find $\phi(18)$, we start with a list of all numbers between 0 and 17:

0 1 2 3 4 5 6 7 8 9 10 11 12 13 14 15 16 17.

Since $18 = 2 \cdot 3^2$, we want to cross off first the even numbers, then the numbers divisible by 3. Start with the evens:

~~0~~ 1 ~~2~~ 3 ~~4~~ 5 ~~6~~ 7 ~~8~~ 9 ~~10~~ 11 ~~12~~ 13 ~~14~~ 15 ~~16~~ ~~17~~

As you might expect, exactly half, or nine, of the 18 numbers are crossed off; half are left. Now we cross off those divisible by 3:

~~0~~ 1 ~~2~~ ~~3~~ ~~4~~ 5 ~~6~~ 7 ~~8~~ ~~9~~ ~~10~~ 11 ~~12~~ 13 ~~14~~ ~~15~~ ~~16~~ 17

Again, of the 18 numbers we started with, exactly 1/3, or six, receive a slash at this stage. What's more, if we look just at the nine numbers left after the first stage, we see that the same fraction 1/3 of them get crossed out at the second stage; likewise, of the nine left after the first step, 2/3, or six, are left after the second. We see in other words that

$$\phi(18) = 18 \cdot \frac{1}{2} \cdot \frac{2}{3} = 6.$$

In fact, exactly the same pattern holds in general. If we're given any number n and asked to find $\phi(n)$, we can just imagine going through the sieve process to find $\phi(n)$—we don't have to actually do it. First, we'd make a list of the n numbers between 0 and $n - 1$. Next, we'd figure out what primes divide n; we can call them p, q, r, and so on. Then we cross off the numbers on our list divisible by p; this involves crossing off exactly $1/p$ of the numbers, so after this stage there are

$$n \cdot \left(1 - \frac{1}{p}\right) = n \cdot \frac{p-1}{p}$$

numbers left. The next step is to cross off the numbers divisible by q, and the key observation here is that this involves getting rid of exactly $1/q$ *of the numbers left after the first stage*—in other words, the number left after the second pass will be

$$n \cdot \frac{p-1}{p} \cdot \frac{q-1}{q}.$$

The process continues in this way: at the third stage, we cross off exactly $1/r$ of the numbers left, so the number remaining after the third is

$$n \cdot \frac{p-1}{p} \cdot \frac{q-1}{q} \cdot \frac{r-1}{r}$$

and so on. Imagining how this is going to go, we're led to the following

Simple Recipe for $\phi(n)$

- First, find the prime factorization of n, and make a list of the primes dividing n.
- Then start with the number n, and for each prime p on your list multiply by $\frac{p-1}{p}$.

Here's an example: suppose we want to find $\phi(84)$. We first write the prime factorization of 84:

$$84 = 2^2 \cdot 3 \cdot 7$$

so the primes dividing 84 are 2, 3 and 7. According to our recipe, then, we have

$$\phi(84) = 84 \cdot \frac{1}{2} \cdot \frac{2}{3} \cdot \frac{6}{7} = 24.$$

Now it's your turn.

Exercise 13.2.3 Find the following:

1. $\phi(55)$
2. $\phi(128)$
3. $\phi(90)$
4. $\phi(89)$
5. $\phi(105)$

Exercise 13.2.4 Verify that the values of $\phi(n)$ you found in Exercise 13.2.1 agree with the recipe.

Exercise 13.2.5

1. How many numbers between 0 and 83 are relatively prime to 84?
2. How many numbers between 0 and 167 are relatively prime to 84?
3. How many numbers between 0 and 463 are relatively prime to 2?

At this point you may have one qualm. (Or maybe you have several; but we're only going to deal with one right now.) As simple as this recipe is, it does require one ingredient that may in some cases be hard to come by: in order to find $\phi(n)$, we have to know the prime factorization of n. And this, you may remember from the last section, is not necessarily easy to find when n is a large number. Of course, there are other methods of calculating $\phi(n)$—for example, there's the brute force method of just listing the numbers from 1 to n and crossing off those with a common factor—but these are too slow to be of any real use.

This seems to be an intrinsic difficulty: there are no known simple recipes for calculating $\phi(n)$ that don't require that we know the prime factorization of n. In fact, as we'll see in the final part of the book, this turns out to be the essential basis of the codes that we use to transmit secure information over the Internet. To put it another way, the safe transmission of electronic information rests on the presumption that you can't figure out a way to calculate $\phi(n)$, for large n, in a reasonably short time.

13.3 Why Does this Work?

You may also be a little uneasy at the idea of looking at two examples and extrapolating from them a general formula. We know that the recipe we gave above for $\phi(n)$ works for $n = 15$ and $n = 18$, and we can also check it for the other values of n for which we've calculated $\phi(n)$ already. But does that mean it works every time?

Actually, you may not be as uneasy as we are. Mathematicians are fairly unique in their desire to see every assertion they make and use proved, and also in their refusal to accept observed behavior as proof. A mathematician will see a pattern repeated a billion times and will not consider it an established fact that the pattern holds in general.[1] Most people would be more than happy to consider the issue settled a lot sooner than that. You're probably not all that upset by the leap of faith we made in the last section—unless you're a mathematician, in which case you're probably livid. And, in our defense, there really are patterns in mathematics that hold in the first thousand or million instances, but are false in general.

What we're saying, in other words, is that the inclusion of the following argument, which tries to demonstrate that the simple recipe given above for $\phi(n)$ always works, may be more for our sake than for yours.

That said, let's start by looking at the second stage of the modified sieve process and seeing why the recipe works there. To say that we're at the second stage means that we've started with a number n, and identified two primes p and q dividing n. We've made a list of the numbers between 0 and $n-1$, and crossed off the numbers divisible by p; that is, there are $\frac{n}{p}$ numbers crossed off and there are

$$n \cdot \frac{p-1}{p}$$

numbers left.

Now we cross off the numbers divisible by q. Since q divides n, this involves crossing exactly n/q of the original n numbers. But how many of those are repeats, that is, numbers already crossed off? Well, the repeats are exactly the numbers divisible by both p and q; and as we saw in the original chapter of this part of the book, since $\gcd(p, q) = 1$, *the numbers divisible by both p and q are just the numbers divisible by pq*. The number n is divisible by pq, so the number of numbers crossed off twice is thus exactly n/pq. We have, in other words

$$\text{Number of numbers we start with} = n$$

$$\text{Number of numbers crossed off in the first step} = \frac{n}{p}$$

$$\text{Number of numbers crossed off in the second step} = \frac{n}{q}$$

and

$$\text{Number of numbers crossed off in both steps} = \frac{n}{pq}$$

Now we can apply the subtraction principle, which tells us that the number of numbers left is

$$\begin{aligned}
\text{Number of numbers left} &= n - \frac{n}{p} - \frac{n}{q} + \frac{n}{pq} \\
&= n\left(1 - \frac{1}{p} - \frac{1}{q} + \frac{1}{pq}\right) \\
&= n\left(1 - \frac{1}{p}\right)\left(1 - \frac{1}{q}\right) \\
&= n \cdot \frac{p-1}{p} \cdot \frac{q-1}{q}.
\end{aligned}$$

[1]This is not an exaggeration. One of the most famous open problems in mathematics is the *Riemann hypothesis*, which asserts that the solutions of a certain equation all lie on a line. This has been verified directly for the first 1,000,000,000 solutions of the equation, but it's not considered an established fact.

In other words, at the second stage we lop off exactly $1/q$ of the remaining numbers, as we said.

The same sort of analysis can be carried out to see that at each stage, when we cross off the numbers divisible by a prime r dividing n, we cross off not just one out of every r of the original numbers, but exactly one out of every r of the numbers remaining at that stage—that is, the simple recipe we gave for $\phi(n)$ works in general.

13.4 Odds and Ends

If we had any self-restraint whatsoever, we'd stop this chapter right here. But we don't. And there is one more fact that is so mind-bogglingly and perversely wonderful that we can't not tell you about it.

To start with, let's go back to the question we raised at the beginning of this chapter: for a particular number n, what proportion of all numbers are relatively prime to n? In other words, what are the odds that a number chosen at random will be relatively prime to n?

The first thing to see is that we have (despite appearances) already answered that question. It may seem that we answered only a very limited version of the question: we asked, "of the n numbers between 0 and $n - 1$, how many are relatively prime to n?" But if we look a little closer, we see that we have in fact answered the original question.

The point is, if a number a is relatively prime to n, then so is $a+n$; if a is not, then $a+n$ isn't either. What this means is that if we look among the next n numbers after $n - 1$ (that is, the numbers between n and $2n - 1$) we'll see exactly the same pattern of numbers relatively prime to n and not as in the numbers from 0 to $n - 1$. For example, when n was 15 we saw that the numbers between 0 and 14 that were relatively prime to 15 are exactly the uncrossed numbers on the list

~~0~~ 1 2 ~~3~~ 4 ~~5~~ ~~6~~ 7 8 ~~9~~ ~~10~~ 11 ~~12~~ 13 14

Now suppose we want to extend the list through another 15 numbers, that is, up to 29. Writing the next 15 numbers under the first 15, we have

~~0~~ 1 2 ~~3~~ 4 ~~5~~ ~~6~~ 7 8 ~~9~~ ~~10~~ 11 ~~12~~ 13 14
~~15~~ 16 17 ~~18~~ 19 ~~20~~ ~~21~~ 22 23 ~~24~~ ~~25~~ 26 ~~27~~ 28 29

and we see that the pattern is the same. In particular, there are exactly as many numbers between 15 and 29 relatively prime to 15 as there are between 0 and 14; and there'll be the same number in the next group of 15, and so on. What this means is that it makes sense to say that, on average, 8 out of 15 numbers are relatively prime to 15. Similarly, we can say in general that

The fraction of all numbers that are relatively prime to n is $\frac{\phi(n)}{n}$.

Now we're going to talk about a problem whose answer we'll only be able to state, not to justify. But it's a natural question to ask, and the answer is startling. If we pick a number n then we've seen how to figure out the percentage of all numbers a that are relatively prime to it—for example, if $n = 15$, then the odds that a number a picked at random will be relatively prime to 15 will be $8/15$. But what if we just pick *two* numbers a and b at random—what are the odds that they're relatively prime to each other, in other words that $\gcd(a, b) = 1$? Equivalently, what are the odds that we can express 1 as a combination of two random numbers a and b?

It's a tricky question, in part because it's not immediately clear that it makes sense. Let's try to make sense out of it, though, by imagining a sort of "thought experiment." Suppose we made a list of all pairs of one-digit numbers (0 to 9), and figured out, of the 100 such pairs, what fraction were relatively prime; suppose we called this fraction x_1. Now say we did the same thing for pairs of 2-digit numbers—in other words, of the 10,000 pairs of numbers between 0 and 99 we figured out what fraction are relatively prime, and called that fraction x_2. Then suppose we did the same for three-digit numbers, and called the resulting fraction x_3, and so on. We would then ask: as we increase the number of digits, how do the numbers $x_1, x_2, x_3, \ldots$ behave? Do they approach 0, or 1, or some number in between? If in fact they do approach a fixed number, we can say that that number represents the fraction of all pairs of numbers that are relatively prime—in other words, the odds that a randomly selected pair of large numbers will be relatively prime.

The answer is that the numbers $x_1, x_2, x_3, \ldots$ do approach a definite number, and a remarkable number at that. We will just state the final result.

> The fraction of all pairs of numbers that are relatively prime approaches the number $\dfrac{6}{\pi^2}$.

Now, $6/\pi^2$ is a real number, with decimal expansion .6079271 ...—in particular, it's a little larger than three-fifths. This may seem a little counter-intuitive: what we're saying is that, if you pick a couple of, say, 50-digit numbers at random, the odds are actually in your favor that you can express 1 as a combination of the two.

But the truly remarkable thing about the number is the factor π^2. What's that doing there? For most of us, the number π represents the ratio between the circumference and diameter of a circle; it also appears in formulas for the area of a circle and the surface area and volume of a sphere. Why on earth should it appear here in the answer to a problem about relatively prime pairs of numbers? This is just one example of the kind of mysteries in mathematics that make it such a fascinating subject.

Title page of the first English translation of Euclid's *Elements*, published in London in 1570.

whole number B A, *wherfore it alſo meaſureth that which remayneth, namely, the number* F A *(by the 4. cōmon ſentence of the ſeuenth). But the number* A F *meaſureth the number* D G, *wherfore* E *alſo meaſureth* D G. *And it meaſureth alſo the whole* D C, *wherfore it alſo meaſureth that which remayneth, namely, the number* G C *(by the ſame common ſentence): but* G C *meaſureth the number* F H, *wherfore alſo* E *meaſureth* F H, *and it meaſureth the whole number* F A, *wherfore (by the former common ſentence) it alſo meaſureth that which remayneth* H A, *which is vnitie, it ſelfe being a number, which is impoſſible. Wherfore no number doth meaſure the numbers* A B *and* C D, *wherfore the numbers* A B *and* C D *are prime numbers the one to the other: which was required to be proued.*

The conuerſe of this propoſition after *Campane.*

The conuerſe of this propoſition.

And if the two numbers, namely A B and C D be prime the one to the other. Then the leſſe being continually taken from the greater there ſhalbe no ſtay of that ſuſtraction, till that you come to vnitie. For if in the continuall ſubſtraction there be a ſtay before you come to vnitie. Suppoſe that H A be the number whereat the ſtay is made, which alſo being ſubtrahed out of G C leaueth nothing. Wherfore H A meaſureth G C wherfore alſo it meaſureth F H by the 5. common ſentence of the ſeuenth. And for as much as it alſo meaſureth it ſelfe, therefore it alſo meaſureth the whole A F by the ſixth common ſentence of the ſeuenth, wherfore alſo it meaſureth D G by the 5. common ſentence. But it is before proued that it meaſureth G C, wherfore it meaſureth the whole C D, by the ſixth common ſentence of the ſeuenth: wherfore alſo it meaſureth B F by the 5. common ſentence of the ſeuenth. And it is alſo proued that it meaſureth F A, wherfore alſo it meaſureth the whole number A B by the ſixth common ſentence of the ſeuenth. Now for as much as the number H A meaſureth the numbers A B and C D, therfore the numbers AB and CD are numbers compoſed: wherfore they are not prime the one to the other: which is contrary to the ſuppoſition.

D . . . G . . C
B . . . . . F . . H . A

How to know whether two numbers geuen be prime the one to the other.

And by this propoſition if there be two numbers geuen. It is eaſy to finde out, whether they be prime the one to the other or no. For if by ſuch continual ſubſtraction of the leſſe from the greater, you come at the length to vnitie. Then are thoſe numbers geuen prime the one to the other. But if there be a ſtay before you come to vnitie, then are the numbers geuen, numbers compoſed the one to the other.

¶ *The 1. Probleme.* *The 2. Propoſition.*

Two numbers being geuen not prime the one to the other, to finde out their greateſt common meaſure.

Two caſes in this probleme. The firſt caſe.

S*Vppoſe the two numbers geuen not prime the one to the other to be* A B *and* C D. *It is required to finde out the greateſt common meaſure of the ſaid numbers* A B *and* C D. *Now the number* C D *either meaſureth the number* A B *or not. If* C D *meaſure* A B *it alſo meaſureth it ſelfe. Wherfore* C D *is a common meaſure to the numbers* C D *and* A B. *And it is manifeſt alſo that it is the greateſt common meaſure: for there is no number greater then* C D *that will meaſure* C D.

A . . . . B
C . . . . D

The ſecond caſe.

But if C D *do not meaſure* A B, *then if of the numbers* A B *and* C D, *the leſſe be continually taken away from the greater, there will before you come to vnitie, be left a number, which will meaſure the number going before (by the 1. of the ſeuenth). For if there ſhould not, then ſhould the numbers* A B *and* C D *be prime the one to the other, which is contrary to the ſuppoſition. Let the ſayd number left by the continuall ſubſtraction of the leſſe number out of the greater be* F C. *So that let the number* C D *meaſuring* A B, *and ſubtrahed out of it as often as you can leue a leſſe number then it ſelfe, namely* A E. *And let* A E *meaſuring* C D, *and ſubtrahed out of it*

A . . . . E . . . . . . B
C . . F . . . . D

as

This page and next: the description of the Euclidean algorithm in Euclid's *Elements.*

as often as you can leaue a lesse then it selfe namely, C F. And suppose that C F do so measure A E that there remayne nothing. Then I say that C F is a common measure to the numbers A B and C D. For forasmuch as C F measureth AE, and A E measureth D F, therefore C F also measureth D F. (by the fifth common sentence of the seuenth) and it likewise measureth it selfe, wherfore it also measureth the whole C D (by the sixth common sentence of the seuenth) but C D measureth B E, wherefore C F also measureth B E (by the fifte common sentence of the seuenth). And it measureth also E A: wherefore it also measureth the whole B A (by the sixth common sentence of the seuenth): and it also measureth C D as we haue before proued: wherefore the number C F measureth the numbers AB & C D: wherfore the number C F is a commo measure to the numbers A B & C D.

Demonstratiõ of the second case.

That C F is a common measure to the numbers A B and C D.

I say also that it is the greatest common measure. For if C F be not the greatest commõ measure to A B and C D, let there be a number greater then C F, which measureth A B and C D: which let be G. And forasmuch as G measureth C D, and C D measureth B E, therefore G also measureth B E (by the fift common sentence of the seuenth). And it measureth the whole A B, wherefore also it measureth the residue, namely, A E (by the 4. common sentence of the seuenth). But A E measureth D F, wherefore G also measureth D F (by the foresayd 5. common sentence of the seuenth). And it measureth the whole C D. Wherefore it also measureth the residue F C: namely, the greater number the lesse: which is impossible. No number therefore greater then C F shall measure those numbers A B and C D: wherefore C F is the greatest common measure to AB and CD: which was required to be done.

A . . . . E . . . . . . B
G . . .
C . . F . . . . D

That C F is the greatest common measure to A B and C D.

Corrolary.

Hereby it is manifest, that if a number measure two numbers it shall also measure their greatest common measure. For if it measure the whole & the part taken away, it shall alwayes measure the residue also, which residue is at the length, the greatest common measure of the two numbers geuen.

¶ The 2. Probleme. Th 3. Proposition.

Thre numbers being geuẽ, not prime the one to the other: to finde out their greatest common measure.

Suppose the three numbers geuen not prime the one to the other to be A, B, C. Now it is required vnto the sayd numbers A, B, C to finde out the greatest common measure. Take the greatest common measure of the two numbers A and B (by the 2 of the seuenth) which let be D: which number D either measureth the number C or not.

A . . . .
B . . . . . .
C . . . .
D . .
E . . .

First let D measure C. And it also measureth the numbers A and B, wherfore D measureth the numbers A, B, C. Wherefore D is a common measure vnto the numbers A, B, C. Then I say also, that it is the greatest common measure vnto them. For if D be not the greatest common measure vnto the numbers A, B, C, let some number greater then D measure the numbers A, B, C. And let the same number be E. Now forasmuch as E measureth the numbers A, B, C, it measureth also the numbers A, B. Wherefore it measureth also the

Two cases in this Proposition. The first case.

PART III

Modular Arithmetic

Chapter 14

What Is a Number?

Here's the thing: the part of the book we're embarking on, an introduction to modular arithmetic, is completely down-to-earth. As you'll see, it's a fascinating game played according to very explicit and easily understood rules. But to set the stage for this, we want to talk a little bit about what we mean by the word "number," at least to the extent of getting you to understand that there really is an issue. Some of the great philosophers of the past century have grappled seriously with this question, so we don't expect to resolve it here; but we do want to get you thinking about your own experience with the concept.

14.1 Pedagogy Recapitulates History

One thing to realize is that while the term "number" may seem self-explanatory, it has meant different things to people at different times in our history, and will probably mean something different to people in the future. The best way to appreciate this may be to think back to your own childhood. For most of us, when we first encounter the notion, a number is what we would now call a *counting number*: 1, 2, 3, and so on; and the first thing we learn is how to add and multiply them. We also learn how to subtract and divide, but these operations aren't always possible. For example, when we subtract 5 from 2, we leave the realm of counting numbers, as we do when we divide 3 by 7.

Partly in response to this, we then enlarge our notion of number. First, we introduce the notion of negative numbers, which is already a good bit less concrete than counting numbers: you can demonstrate to a child what 3 means by pointing to three apples or three chairs, but it's a lot more difficult to explain what -3 is. One way to do this is to say that -3 is the answer to the question "what do you get when you take 5 away from 2?" For most of us, this is the first step into the world of abstraction. It's not a major stumbling block, as most of us seem comfortable with the sort of practical definition "-3 is the number that, when you add it to 5, gives 2." We thus arrive at what are called *whole numbers*, or *integers*: $\{\ldots, -2, -1, 0, 1, 2, \ldots\}$.

Note that there is a potential problem inherent in introducing new numbers so as to permit an operation (like subtraction) to be carried out in general. What becomes of the other operations (like multiplication)? For example, if we define -3 as "the number that, when you add it to 5, gives 2," how do we know what we get when we multiply -3 by 2? If we define -2 analogously, how do we know what we get when we multiply -3 by -2? In fact, we'll see the answer to this in Section 14.3 below.

Once we've enlarged our concept of number to include negative numbers, we can always carry out subtractions, but division remains problematic. Accordingly, for most of us the next step is to introduce fractions: that is, ratios of whole numbers. We introduce symbols like $\frac{2}{3}$ and $\frac{3}{7}$, and we learn how to add and multiply them.

Once more, these are a little more abstract than the counting numbers: it's true that we can illustrate fractions by pointing to physical objects, but can we say *exactly* what we mean by $\frac{5}{11}$ of a cookie? If you stop and think about it—and certainly if you've ever had the experience of explaining these things to a child—you realize that it's not so easy to say precisely what we mean by a fraction. Nonetheless, most of us get along here, thinking of $\frac{2}{3}$—if we think of it at all—as the number that, when you multiply by 3, you get 2. And learning the rules for adding, multiplying, subtracting and dividing these numbers reassures us: as long as we can carry out basic arithmetic operations on them in a consistent way, who cares whether we can define them rigorously?

The bottom line is that fractions suffice for all economic, and even for all scientific, calculations in the real world. However, they do not suffice for even the simplest measurements in abstract geometry, as the Pythagoreans discovered over 2500 years ago.

14.2 $\sqrt{2}$ Is Still Not a Fraction

Although Pythagoras looms large in the popular histories of mathematics, we know almost nothing about his history. In some sense, he is as mythical as his contemporary, Prometheus. There was a sect, called the Pythagoreans, whose results were all attributed to their leader. One of their most famous results was the Pythagorean theorem, which we all encounter in high school. This says that if you have a right triangle, the square of the length of the hypotenuse is the sum of the squares of the two sides. The favorite illustration of this is the 3-4-5 right triangle: because

$$3^2 + 4^2 = 9 + 16 = 25 = 5^2,$$

a right triangle with sides of length 3 and 4 will have hypotenuse of length 5. Another example is the 5-12-13 right triangle, since $25 + 144 = 169$. The triples of numbers (a, b, c) with

$$a^2 + b^2 = c^2$$

are called *Pythagorean triples*. There are an infinite number of them; a procedure for producing all Pythagorean triples was known to the Greeks, and perhaps even to the Babylonians.

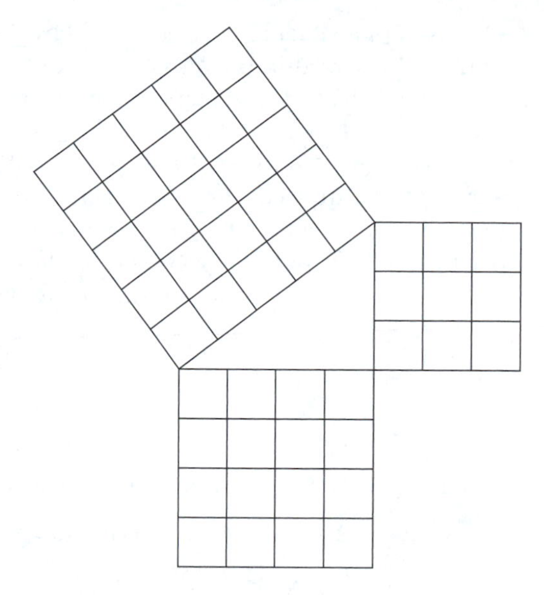

Of course, the sides of a right triangle don't *all* have to be whole numbers. Suppose that we have a right triangle whose sides are both of length 1? What will be the length of its hypotenuse?

The answer, by the Pythagorean theorem, is that the length of the hypotenuse squared will be $1^2 + 1^2 = 2$—in other words, the length will be the square root of 2. But as we've seen in Section 12.6, *there is no fraction whose square is* 2. Hence, we have a simply defined length in abstract geometry, which is not given by a fraction!

All of which means that it's time to enlarge our number system again, to include things like $\sqrt{2}$. For most of us today, this is not too difficult: we typically encounter numbers like $\sqrt{2}$ for the first time in junior high school, a setting where we expect to be routinely baffled. We learn rules for working with numbers like $\sqrt{2}$—to add and multiply numbers involving such symbols—even if once more we would be hard-pressed to explain precisely what the symbols we're manipulating represent.

It was a lot more difficult in ancient Greece and Egypt, where the existence of lengths that were not in fractional ratio, like the side and hypotenuse of an isosceles right triangle, caused a myriad of problems. In the ancient world, feelings ran pretty high on this issue. For example, the first female mathematician known to history was a neo-Platonist named Hypatia, who lived in Alexandria in the fifth century A.D. The neo-Platonists believed (among other things) in the ability of the human mind to conceive of numbers beyond those dealt with at the time; Hypatia herself was put to death for her beliefs by an outraged mob.[1] In the end, these ratios, called *incommensurables*, necessitated a rewriting of much of the mathematics known at

[1] In all honesty, it's unlikely that Hypatia's belief in the existence of a square root of 2 led directly to her death. In fact, Hypatia was outspoken on a number of issues; she was, for example, on record as saying, "Reserve your right to think, for even to think wrongly is better than not to think at all" and "To teach superstitions as truth is a most terrible thing." What mob wouldn't be incensed? For more about Hypatia, see the Encyclopedia Britannica article, at `http://www.britannica.com/eb/article?eu=42725&tocid=0`.

the time. The tools to treat them were developed by the great classical mathematician Eudoxus, and are presented in Euclid's treatise on geometry.

The Greek mathematicians thought of these new numbers in geometric terms: they were defined as lengths, or ratios of lengths, of geometric objects. A more modern mathematician, on the other hand, might argue for them in algebraic terms, as follows. Suppose we introduced a new number, denoted by the symbol $\sqrt{2}$, with the property that when you multiply it by itself you get 2. We could consider as numbers any combinations of our old numbers and this new one: our numbers would look like $1+3\sqrt{2}$, or $2+\sqrt{2}$. We could still add them and multiply them: for example,

$$(1+3\sqrt{2})+(2+\sqrt{2})=3+4\sqrt{2}$$

and

$$\begin{aligned}(1+3\sqrt{2}) &\times (2+\sqrt{2})\\ &= (1\times 2)+(1\times\sqrt{2})+(3\sqrt{2}\times 2)+(3\sqrt{2}\times\sqrt{2})\\ &= 2+\sqrt{2}+6\sqrt{2}+(3\sqrt{2}\times\sqrt{2})\\ &= 2+\sqrt{2}+6\sqrt{2}+6\\ &= 8+7\sqrt{2}.\end{aligned}$$

In other words, we could do everything with these new numbers that we're currently able to do with fractions, and we'd have a square root of 2.

The trade-off here is that, in exchange for the additional flexibility, our notion of "number" has again become more abstract. If you were to try to explain to an 8-year old that "$\sqrt{2}$ is the number that, when you multiply it by itself, you get 2," you would be told (quite reasonably, in our view) that "there is no such number."

The next major development took place in the 16th–18th centuries with the introduction of a new system of numbers, the *decimal*, or *real* numbers. In practice, most of us would be hard-pressed to say exactly what we meant by an infinite decimal. We know what we mean by a finite decimal: the number 3.141 means

$$3+\frac{1}{10}+\frac{4}{100}+\frac{1}{1{,}000} \quad \text{or in other words} \quad \frac{3{,}141}{10{,}000};$$

the decimal 3.141, in other words, represents simply another way to write a fraction. But what do we mean by an infinite decimal, like

$$\pi = 3.14159265358979323846264338327 9\ldots?$$

We know that it's bigger than $\frac{3{,}141}{10{,}000}$, but less than $\frac{3{,}142}{10{,}000}$; it's bigger than $\frac{3{,}141{,}592}{1{,}000{,}000}$ but less than $\frac{3{,}141{,}593}{1{,}000{,}000}$, and so on. In fact, that's exactly what we mean by an infinite decimal: it's the number that satisfies a series of ever-finer approximations like this. But why should there be any such number, and why should there be only one? For example, are the decimal numbers $1.9999999\ldots$ and $2.0000000\ldots$ equal?

In fact, these days we learn early enough in our schooling to manipulate decimals that we accept them as numbers, even if we are unable to define precisely the concept of an infinite decimal. Our point here is, once again, that our notion of what is a number is a functional, rather than an abstract one: a number is something we know how to add and multiply.

Nor is this the end of the line. Mathematicians proposed, for algebraic purposes, enlarging our decimal number system still further, to include a square root of -1, denoted by the letter i. We can write numbers as combinations of our old numbers and this new one: our numbers would now look like $1+3i$, or $2+i$. We could still add and multiply them: for example,

$$(1+3i)+(2+i)=3+4i$$

and

$$\begin{aligned}(1+3i)&\times(2+i)\\&=(1\times 2)+(1\times i)+(3i\times 2)+(3i\times i)\\&=2+i+6i+3i\times i\\&=2+i+6i-3\\&=-1+7i.\end{aligned}$$

In other words, we could do everything with these new numbers that we're currently able to do with fractions, and we'd have a square root of -1.

The system of numbers we get if we start with decimals and tack on the square root of -1 (called the *complex numbers*) is ubiquitous in math, science and engineering, but hasn't filtered down to the consciousness of the average person. Which may be just as well: at least no one has been dismembered recently for professing a belief in the existence of complex numbers.

14.3 What Is a Number?

The point of the preceding section is simply to show that there are different possible notions of what a number is. In the end, what we're proposing is a functional, rather than a teleological, definition.

To begin with, let's say what a number system is. This starts out simply enough: *a number system is a collection of symbols that we can add and multiply*. Of course, there have to be some rules. Most often, we study number systems that satisfy these axioms:

- *The order of the terms in a sum or product shouldn't matter*: if a and b are numbers, the sum $a+b$ should be the same as $b+a$, and the same for products: ab is the same as ba.
- *The order of combination shouldn't matter*: if a, b and c are numbers, if we add a to b and then add the result to c, we should get the same result as if we had added b to c and then added a to that. Again, the same should be true for products. In math,

  $$(a+b)+c=a+(b+c) \qquad \text{and} \qquad (a\times b)\times c=a\times(b\times c)$$

- *The distributive law*: If we multiply a number a by the sum of two numbers b and c, we should get the same result as if we multiplied a by each of b and c and added up the results. In math,

  $$a(b+c)=ab+ac.$$

Why do we insist on these axioms? Well, to some extent it's just tradition; we could, for example, define "number system" even more broadly by dropping one, and in fact mathematicians have studied such generalized number systems as well.

But they also serve a purpose, in that when we want to enlarge our number system they allow us to extend the basic operations unambiguously. For example, suppose we start with the counting numbers $\{1, 2, 3, \ldots\}$ and want to introduce a negative number like -3, defined (as in the first section of this chapter) as "the number that, when you add it to 5, you get 2." It may not be immediately clear then how to multiply -3 by, say, 2; but the distributive law forces the answer on us: since $-3 + 5 = 2$, we can apply the distributive law and see that

$$\begin{aligned} 2 \times 2 &= (-3 + 5) \times 2 \\ &= (-3 \times 2) + (5 \times 2) \\ &= (-3 \times 2) + 10. \end{aligned}$$

Thus we see that $4 = (-3 \times 2) + 10$, and hence that $-3 \times 2 = -6$.

To sum up, we've seen several possible number systems. In the order in which they've appeared—both in history, and (for most of us) in the development of our knowledge of numbers—they are

- *counting numbers*: $\{1, 2, 3, \ldots\}$. In mathematics, these are called the *natural numbers*, and are usually denoted $\mathbb{N}$.
- *whole numbers*: $\{\ldots, -2, -1, 0, 1, 2, \ldots\}$. In mathematics, these are called the *integers*, and are usually denoted $\mathbb{Z}$.
- *fractions*: ratios $\{a/b\}$ of whole numbers. In mathematics, these are called the *rational numbers*, and are usually denoted $\mathbb{Q}$.
- *decimals*: possibly infinite decimal numbers, like $\pi = 3.14159\ldots$. In mathematics, these are called the *real numbers*, and are usually denoted $\mathbb{R}$.
- *complex numbers*: like $\pi + 3i$. The collection of these numbers is usually denoted $\mathbb{C}$.

Each of these systems is included in the next, and each satisfies all the axioms we have proposed.

The main point of this chapter, though, is that *there is no a priori notion of what a number should be*. What we're saying, in other words, is that a number is just something you can add and multiply with other numbers. To illustrate the point, we'll show you now a number system you've (probably) never seen before.

To start with, this number system will have only 4 numbers in it, represented by the symbols $\spadesuit$, $\heartsuit$, $\diamondsuit$ and $\clubsuit$. Next, we have to specify how to add two of these symbols and how to multiply them. We'll do that by giving you an addition table and a multiplication table:

$+$	$\spadesuit$	$\heartsuit$	$\diamondsuit$	$\clubsuit$
$\spadesuit$	$\spadesuit$	$\heartsuit$	$\diamondsuit$	$\clubsuit$
$\heartsuit$	$\heartsuit$	$\spadesuit$	$\clubsuit$	$\diamondsuit$
$\diamondsuit$	$\diamondsuit$	$\clubsuit$	$\spadesuit$	$\heartsuit$
$\clubsuit$	$\clubsuit$	$\diamondsuit$	$\heartsuit$	$\spadesuit$

$\times$	$\spadesuit$	$\heartsuit$	$\diamondsuit$	$\clubsuit$
$\spadesuit$	$\spadesuit$	$\spadesuit$	$\spadesuit$	$\spadesuit$
$\heartsuit$	$\spadesuit$	$\heartsuit$	$\diamondsuit$	$\clubsuit$
$\diamondsuit$	$\spadesuit$	$\diamondsuit$	$\clubsuit$	$\heartsuit$
$\clubsuit$	$\spadesuit$	$\clubsuit$	$\heartsuit$	$\diamondsuit$

These tables tell us how to perform the basic operations on these four "numbers": for example, we can look in the addition table and see that $\diamondsuit + \heartsuit = \clubsuit$; and we can look in the multiplication table and see that $\diamondsuit \times \clubsuit = \heartsuit$. Of course, we do have to verify that the rules given in the tables satisfy the axioms; we'll ask you to do a couple such verifications in Exercise 14.3.1 below, just to give you an idea of what's involved.

Now, you might object that $\spadesuit$, $\heartsuit$, $\diamondsuit$ and $\clubsuit$ aren't numbers. That's the whole point! It's true, suit symbols don't count anything. But then again, the square root of 2 and the complex number $i = \sqrt{-1}$ aren't numbers in that sense either; and mathematicians, scientists and engineers work with them every day. In this sense, the collection of symbols $\spadesuit$, $\heartsuit$, $\diamondsuit$ and $\clubsuit$, together with the rules for addition and multiplication expressed in the tables above, comprise a perfectly legitimate number system. In the next chapter, we'll see how to construct a whole series of new number systems that exist wholly outside of our normal experience, and which we'll study in some detail in the following chapters.

Exercise 14.3.1 Using the tables above, verify that the following equalities hold:

1. $(\diamondsuit + \heartsuit) + \clubsuit = \diamondsuit + (\heartsuit + \clubsuit)$
2. $(\diamondsuit + \clubsuit) \times \heartsuit = (\diamondsuit \times \heartsuit) + (\clubsuit \times \heartsuit)$

14.4 Mathematicians in Outer Space

Imagine for a moment a geologist. She grows up studying mountains all over the world, and comes to some understanding of the laws that govern their creation: how rocks behave under pressure and temperature changes, and what formations result. But in the end, all the mountains she's studied are made up of the elements prevalent on this planet, under conditions that may be peculiar to this planet. What would it be worth to that geologist to be able to visit and study a mountain on another planet?

Or imagine a biologist who's studied species throughout the globe, and has formulated theories about how life forms and evolves. How can he test those theories? The scientific method requires that we formulate hypotheses and then test them by gathering new data, but where is the new data to come from? Again, what would it be worth to that biologist to study life forms from other planets?

Well, mathematicians are in a similar situation. We grow up studying the properties of the basic number systems we've described here, and trying to formulate some understanding of how they work. But unlike the geologist and the biologist, *we get to go into outer space: we can make up our own number systems and study them.* We can discover for ourselves what aspects of the familiar number system are truly universal, and which are peculiar to it.

We will show, in the following chapter, how to do this: how to set up a collection of new number systems and study their arithmetic. Mathematicians do this primarily for their own edification and amusement, and we'd suggest you take the same attitude: the rules that govern modular arithmetic are easy to learn and fun to play with, and at the end of a few weeks you'll have had a new experience. Remarkably enough, though, there are practical applications of modular arithmetic as well; we'll describe some of those in the last part of the book.

HUTT ONE...
HUTT TWO...
HUTT THREE...
11-25

...POINT 14159265358979323846264
338327950288419716939937510582
09749445923078164062862089986
280348253421170679821480865132
82306647093844609550582231725
359408128 48111745028410270193
852110555964462294...UM...4...UM...
© 2000 Bill Amend / Distributed by Universal Press Syndicate / www.foxtrot.com

WHOSE IDEA WAS IT TO HIKE THE BALL ON PI, AGAIN?
YOURS. NOW KEEP GOING.
AMEND

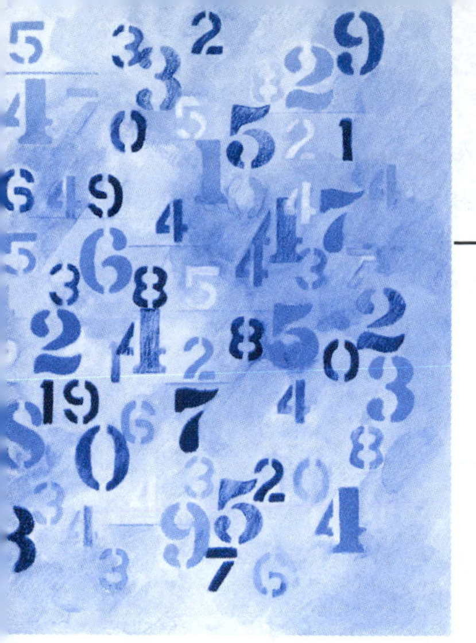

Chapter 15

Modular Arithmetic

The new number systems that we're going to introduce in this chapter and study for the remainder of this part of the book are called, collectively, *modular arithmetic*. We'll start our investigation of modular arithmetic with an example: arithmetic mod 5.

15.1 Arithmetic mod 5

First off: if you're expecting something elaborate and difficult, relax. As we said at the outset of the last chapter, modular arithmetic is completely down-to-earth and follows straightforward rules.

To begin with, in the number system "arithmetic mod 5" there are only five numbers, which we represent by the symbols

$$0 \quad 1 \quad 2 \quad 3 \quad \text{and} \quad 4.$$

The rules for addition are simple enough: we add two of these symbols just as we would in ordinary arithmetic, with the one twist that *if the sum of two numbers exceeds 4, we subtract 5 to arrive at a number between 0 and 4*. Thus $2+1$ is 3, just like always; but to add 3 and 4, for example, we take the ordinary sum 7 and subtract 5 to arrive at 2. Now, since the symbols we're using in arithmetic mod 5 look the same as whole numbers, we need to use a different symbol to indicate equality; we use the symbol $\equiv$, and tack on "(mod 5)" afterward to mean that we are working in this other number system. Thus, we would write the two examples above as

$$2+1 \equiv 3 \pmod 5 \qquad \text{and} \qquad 3+4 \equiv 2 \pmod 5;$$

other examples would be

$$2+3 \equiv 0 \pmod 5 \qquad 3+3 \equiv 1 \pmod 5 \qquad \text{and} \qquad 4+4 \equiv 3 \pmod 5.$$

We put all this information into Table 15-1, an addition table for arithmetic mod 5.

We should probably take a moment here and check that this rule for addition in arithmetic mod 5 really does satisfy the requirements we gave in Section 14.3 (or at least those that involve addition alone; the other ones we'll deal with after we describe multiplication in this number system). It's pretty obvious that when we add

TABLE 15-1 The Addition Table for Arithmetic Mod 5

+	0	1	2	3	4
0	0	1	2	3	4
1	1	2	3	4	0
2	2	3	4	0	1
3	3	4	0	1	2
4	4	0	1	2	3

two numbers the order doesn't matter. As for the second requirement, to see that $(a+b)+c$ yields the same answer as $a+(b+c)$ in arithmetic mod 5—that is, the order of combination doesn't matter—we can certainly check it in specific cases: for example,

$$(3+4)+2 \equiv 2+2 \equiv 4 \pmod 5$$

and

$$3+(4+2) \equiv 3+1 \equiv 4 \pmod 5$$

so we're OK here. If checking a bunch of examples like this doesn't convince you, probably the best way to see that $(a+b)+c \equiv a+(b+c) \pmod 5$ holds for any a, b and c would be to observe that they're both equal to the result we'd get if we simply added a, b and c as ordinary numbers and then took the remainder after division by 5.

Exercise 15.1.1 Verify the following equalities by carrying out the additions specified:

1. $(4+4)+3 \equiv 4+(4+3) \pmod 5$
2. $(1+2)+4 \equiv 1+(2+4) \pmod 5$

15.2 Subtraction

We'll go on to talk about multiplication in a moment, but before we do we want to mention subtraction. The first thing we need to do is to say what we mean by subtraction.

In any number system, what we mean by subtracting one number b from another number a is finding "the number that, when you add b to it, you get a"—if one exists. Note that in counting numbers, you may or may not be able to subtract one number from another: for example, there's no counting number you can add to 7 to get 4, so $4-7$ doesn't exist in this number system. But in whole numbers, fractions or decimals you can always subtract one number from another.

Can we perform subtraction in arithmetic mod 5? To answer this, take another look at the addition table above. The crucial thing to notice here is that *every row of this table contains every number in arithmetic mod 5 exactly once*. What this means, if you think about it, is that *we can always carry out subtraction in arithmetic mod 5.* For example, suppose we want to subtract 4 from 1 in arithmetic mod 5. By definition, $1-4$ is the number that, when you add 4 to it, you get 1. So we look at the

"4" row of the addition table, and spot the 1 in the "2" column. This means that $4 + 2 \equiv 1 \pmod 5$; or, equivalently, $1 - 4 \equiv 2 \pmod 5$. Similarly, you can use the table to verify these other subtractions:

$$2 - 3 \equiv 4 \pmod 5 \qquad 2 - 4 \equiv 3 \pmod 5 \qquad \text{and} \qquad 0 - 4 \equiv 1 \pmod 5.$$

To say it in general: for any numbers a and b in arithmetic mod 5, to find $a - b$ you look in the "b" row of the addition table; you will find an a. If the a appears in the "c" column, then c is the number you add to b to get a, or in other words $a - b \equiv c \pmod 5$.

There is another way of describing the operation of subtraction in arithmetic mod 5 which is perhaps easier to carry out in practice: *we subtract one of these numbers from another just as we would in ordinary arithmetic; and if the result is negative, we add 5 to arrive at a number between 0 and 4.* For example, to do the subtraction $1-4$ that we did above, we subtract 4 from 1 in whole numbers, yielding -3, to which we add 5 to arrive at 2. This may seem like a more straightforward prescription for subtraction (it is). When we get to division, you'll see why we introduced the other way of doing subtraction first.

While we're at it, we should say what we mean by negative numbers in arithmetic mod 5. This is easy: by a negative number like -3 we mean "the number such that, when you add 3 to it, you get 0"—in other words, $0 - 3$. And again we can simply look in our addition table to find the negatives of the numbers in arithmetic mod 5: for example, we see that

$$-1 \equiv 4, \quad -2 \equiv 3, \quad -3 \equiv 2 \quad \text{and} \quad -4 \equiv 1 \quad \pmod 5.$$

Note that we could also have talked about negative numbers first, and then used them to define subtraction: for example, if we know that $-4 \equiv 1 \pmod 5$, then to carry out the subtraction $2 - 4$ in arithmetic mod 5 we can write

$$2 - 4 \equiv 2 + (-4) \equiv 2 + 1 \equiv 3 \pmod 5.$$

Exercise 15.2.1 Evaluate $1 - 3$ in arithmetic mod 5 in two ways:

1. by consulting the addition table above; and
2. by adding -3 to 1 in arithmetic mod 5.

15.3 Multiplication

The basic rule for multiplication in arithmetic mod 5 is very much like addition: if we want to multiply two numbers, we multiply them as whole numbers, and then subtract 5s until we get to a number between 0 and 4. For example, if we want to multiply 2 by 3, we multiply them as regular numbers and get 6; since that's larger than 4 we subtract 5 to get 1, so that

$$2 \times 3 \equiv 1 \pmod 5$$

Similarly, if want to find 3×4 in arithmetic mod 5, we multiply them to get 12, then subtract 5 twice to get 2; thus

$$3 \times 4 \equiv 2 \pmod 5$$

Table 15-2 is the multiplication table for arithmetic mod 5; you should pick a few products at random and check that we have the right entries.

TABLE 15-2 The Multiplication Table for Arithmetic Mod 5

×	0	1	2	3	4
0	0	0	0	0	0
1	0	1	2	3	4
2	0	2	4	1	3
3	0	3	1	4	2
4	0	4	3	2	1

Now that we've defined both addition and multiplication in arithmetic mod 5, and before we take a closer look at some of the fascinating features of these tables, let's take a moment out to check that the basic rules do apply here. As in the case of addition, it's pretty clear to begin with that $a \times b \equiv b \times a \pmod 5$ for any a and b. As for the second rule—that the parenthesization of a triple product doesn't matter—again, we can check it in specific cases. For example, let's multiply out $2 \times 3 \times 4$. We do this two ways:

$$(2 \times 3) \times 4 \equiv 1 \times 4 \equiv 4 \pmod 5;$$

and

$$2 \times (3 \times 4) \equiv 2 \times 2 \equiv 4 \pmod 5;$$

and we see that we do indeed get the same answer either way. As before, if you want to see that $(a \times b) \times c \equiv a \times (b \times c) \pmod 5$ for any three numbers a, b and c, probably the best way would be to convince yourself that they're both equal to the result of taking the product of a, b and c as whole numbers (where we know the parenthesization doesn't matter) and then taking the remainder after division by 5.

We can also use the same argument to convince ourselves that the third requirement for a number system, the distributive law, holds in arithmetic mod 5. But it doesn't hurt to do some examples ...

Exercise 15.3.1 Verify the distributive law in a couple of cases: for example, see that

$$4 \times (3 + 4) \equiv (4 \times 3) + (4 \times 4) \quad \pmod 5$$

by adding and multiplying both sides out.

Exercise 15.3.2 In arithmetic mod 5, multiplying a number by 4 always gives you its negative. For example, $4 \times 3 \equiv -3 \pmod 5$. Why is this?

There are many fascinating aspects of the multiplication table in arithmetic mod 5. For example, note that in every row of the multiplication table except for the top (that is, the "0" row), every number in arithmetic mod 5 appears. What does this mean? (Go back to our discussion of subtraction for a clue.) Meanwhile, here is another:

In any number system, we can talk about square roots: the square root of a number a is the number you multiply by itself to get a—if there is one. In counting numbers and whole numbers, some numbers—1, 4, 9, 16 and so on—have square roots, but most do not. Likewise, in fractions, some numbers have square roots but most do not: based on what we saw when we did unique factorization, if we write a fraction in lowest terms, it will be a square only if the numerator and denominator are each squares in their own right. In the real number system, though, every positive number has a square root—that is, half the nonzero numbers do. And in the complex numbers, every number has a square root (though we won't prove this).

So: which of these number systems is arithmetic mod 5 like in this respect? We can see by consulting the multiplication table once more: of the nonzero numbers 1, 2, 3 and 4, two of them—1 and 4—have square roots. This follows from the formulas:

$$1 \times 1 \equiv 4 \times 4 \equiv 1 \pmod 5 \quad \text{and} \quad 2 \times 2 \equiv 3 \times 3 \equiv 4 \pmod 5$$

On the other hand, two nonzero numbers—2 and 3—do not have square roots. In other words, half the nonzero numbers in arithmetic mod 5 have square roots. In this respect, at least, it seems that arithmetic mod 5 behaves like the system of decimal numbers!

Exercise 15.3.3 Which numbers in arithmetic mod 5 are cubes? Which are 4th powers?

15.4 Other Flavors of Modular Arithmetic: What's So Special About 5?

Absolutely nothing, as far as the material covered so far. We could define number systems called "arithmetic mod 6", "arithmetic mod 7", or mod any number n we want. The basic rules for arithmetic mod n are these:

- the numbers in arithmetic mod n are the whole numbers from 0 up to $n - 1$
- To add two numbers in arithmetic mod n, we add them as whole numbers and subtract n if necessary to bring the answer between 0 and $n - 1$
- To multiply two numbers in arithmetic mod n, we multiply them as whole numbers and subtract n as many times as necessary to bring the answer between 0 and $n - 1$
- More generally, to perform any sequence of sums and products in arithmetic mod n, we perform the sequence of operations in the natural numbers, then take the remainder (between 0 and $n - 1$) after division by n

The number n is called the *modulus* of the number system arithmetic mod n. The language, as well as the symbol $\equiv$ to denote equality in these number systems, comes from Gauss' great book, *Disquisitiones Arithmeticae*, published in 1801. We've reproduced two pages of this groundbreaking work at the end of Part III.

The simplest example of these number systems is, naturally, arithmetic mod 2. Here we have only two numbers, represented by the symbols 0 and 1 (this is really about as small as a number system can get!). Addition is carried out just as in whole numbers, except that $1 + 1$ comes out to be 0 rather than 2. Multiplication is exactly the same, since taking products of 0s and 1s will never yield anything but more 0s and 1s. The rules for addition and multiplication in arithmetic mod 2 are shown in Table 15-3

TABLE 15-3 Tables for Arithmetic Mod 2

+	0	1
0	0	1
1	1	0

x	0	1
0	0	0
1	0	1

For a slightly meatier example, let's look at arithmetic mod 6. In this system, there are exactly six numbers, represented by the symbols 0, 1, 2, 3, 4 and 5. The rules for addition and multiplication are, as always, to carry out the same operations as whole numbers and then throw away 6s until the answer is between 0 and 5. The tables for arithmetic mod 6 are given in Table 15-4.

TABLE 15-4 Tables for Arithmetic Mod 6

+	0	1	2	3	4	5
0	0	1	2	3	4	5
1	1	2	3	4	5	0
2	2	3	4	5	0	1
3	3	4	5	0	1	2
4	4	5	0	1	2	3
5	5	0	1	2	3	4

x	0	1	2	3	4	5
0	0	0	0	0	0	0
1	0	1	2	3	4	5
2	0	2	4	0	2	4
3	0	3	0	3	0	3
4	0	4	2	0	4	2
5	0	5	4	3	2	1

Let's look at these tables, and compare them in particular to the corresponding tables for arithmetic mod 5, which we'll reproduce here.

TABLE 15-5 Tables for Arithmetic Mod 5

+	0	1	2	3	4
0	0	1	2	3	4
1	1	2	3	4	0
2	2	3	4	0	1
3	3	4	0	1	2
4	4	0	1	2	3

x	0	1	2	3	4
0	0	0	0	0	0
1	0	1	2	3	4
2	0	2	4	1	3
3	0	3	1	4	2
4	0	4	3	2	1

Certainly the addition table for arithmetic mod 6 shows many of the same patterns that we saw in the corresponding table for arithmetic mod 5. For one thing, each row is just the row above it shifted to the left by one column, with the number that was in the "0" column in the row above now appearing in the last column (so that for example the 5s appear exactly on the diagonal in the addition table mod 6, just as the 4s appear exactly on the diagonal in the addition table mod 5). In particular, we see that every number appears exactly once in each row, which tells us that we can perform subtraction in this number system.

The multiplication tables, on the other hand, tell a different story. Look first at the multiplication table for arithmetic mod 5. The first thing to notice is that, while the numbers jump around from row to row, each row (except the "0" row) contains each number exactly once. Since the table is symmetric ($a \times b = b \times a$), the same is true of each column, again except the "0" column. In particular, each row except the "0" row contains a 1, which means that for any number other than 0 there's a number you can multiply it by to get 1: for example, $3 \times 2 \equiv 1 \pmod 5$. This feature of the multiplication table will allow us to define division.

The multiplication table for arithmetic mod 6 couldn't be more different. Two of the rows look similar: the "1" row contains the sequence 1, 2, 3, 4, 5 and the last row is the same sequence in reverse. (But if you think about it, the same will be true in every flavor of modular arithmetic; see Exercise 15.4.3 below.) In between, though, things look very different: rows do not always contain 1s; and there are 0s outside of the first row and column, which there weren't in the corresponding table for arithmetic mod 5. For now, think about why this is the case and what it means; we'll answer those questions fully when we get to Chapter 17.

Exercise 15.4.1 Evaluate the following expressions in modular arithmetic:

1. $12 + 13 \pmod{17}$
2. $6 - 13 \pmod{17}$
3. $5 \times 7 \pmod{23}$
4. $9 \times (14 + 19) \pmod{23}$ (Do this in two ways!)

Exercise 15.4.2 Write out the addition and multiplication tables for arithmetic mod 7, and then use them to do the following problems:

1. Find $3-6$ in arithmetic mod 7: that is, find a number x so that $6+x \equiv 3 \pmod 7$.
2. Find $1/3$ in arithmetic mod 7: that is, find a number x so that $3x \equiv 1 \pmod 7$.
3. Find $3/5$ in arithmetic mod 7: that is, find a number x so that $5x \equiv 3 \pmod 7$.
4. Find a square root of 2 in arithmetic mod 7
5. Look at the patterns in these tables: do they have more in common with those for arithmetic mod 5 or arithmetic mod 6?

Exercise 15.4.3 Write out the bottom row of the multiplication table for arithmetic mod 10 and for arithmetic mod 11 (that is, the "9" row and the "10" row, respectively). Does the same pattern hold in arithmetic mod n for any n? Why?

TABLE 15-6 Tables for Arithmetic Mod 3

+	*0*	*1*	*2*
0	0	1	2
1	1	2	0
2	2	0	1

×	*0*	*1*	*2*
0	0	0	0
1	0	1	2
2	0	2	1

15.5 The Fine Print

One warning about a potential source of confusion here. The various number systems we all grew up with—counting numbers, whole numbers, fractions, decimals and so on—are all compatible with one another. That is, each is obtained from the ones before it by including additional numbers, without changing the rules of addition and multiplication for the numbers already in the system: in other words, if $2 + 3 = 5$ in the whole number system, it will still be true in the world of fractions, of real numbers, and so on.

The same is *not* true of the various flavors of arithmetic mod n: they are not compatible with each other, or with any of the standard number systems. This is potentially confusing, because some of the symbols we use to denote numbers in arithmetic mod 7 are the same as the ones we used in arithmetic mod 5 and in ordinary arithmetic. But the rules of addition and multiplication are fundamentally different: we have $4 + 4 \equiv 3 \pmod{5}$, but $4 + 4 \equiv 1 \pmod{7}$.

HERE'S ANOTHER MATH PROBLEM I CAN'T FIGURE OUT. WHAT'S 9+4?
OOH, THAT'S A TRICKY ONE. YOU HAVE TO USE CALCULUS AND IMAGINARY NUMBERS FOR THIS.
IMAGINARY NUMBERS?!
YOU KNOW, ELEVENTEEN, THIRTY-TWELVE, AND ALL THOSE. IT'S A LITTLE CONFUSING AT FIRST.
HOW DID YOU LEARN ALL THIS? YOU'VE NEVER EVEN GONE TO SCHOOL!
INSTINCT. TIGERS ARE BORN WITH IT.
©1988 Universal Press Syndicate
1-6
WATTERSON

Chapter 16

Congruences: Another Way to Look at Modular Arithmetic

16.1 Evens and Odds

Let's start with something familiar: if you add two even numbers, you get an even number. If you add an even and an odd number, you get an odd number; and if you add two odd numbers you get an even number.

Most people also know how odd and even numbers behave under multiplication: if you multiply two numbers, the result will be even if either of the numbers you start with is, while if you multiply two odd numbers you get an odd number. We can collect these facts in Table 16-1.

TABLE 16-1 The Rules for Combining Odd and Even Numbers

+	*even*	*odd*
even	even	odd
odd	odd	even

×	*even*	*odd*
even	even	even
odd	even	odd

Why are these things such common knowledge? Is it that they're truly self-evident, or is there something special in human cognition about the classification of numbers into those divisible by 2 and those not? Well, we're mathematicians, not anthropologists; what we're going to do is examine the phenomena described in Table 16-1, try to give a precise formulation of what's going on, and then see if the same sort of thing can be done with a different number—say, 3—in place of 2.

Before we start, we have to recognize that there's something remarkable going on here. We divide the world of whole numbers into those that are divisible by 2 and those that aren't; that's no big deal. But now what we're saying is that *the type of number we get as the outcome of any addition or multiplication depends only on the types of the numbers we're combining*. That, if you think about it, *is* a big deal, and one worth investigating.

16.2 Threvens and Throdds

Now let's try to do the same thing with 3 in place of 2. In other words, let's divide all numbers into two classes: those that are divisible by 3, and those that aren't. We'll call the first group "threven" and the second "throdd." We then ask: can we predict what type of number—threven or throdd—we get when we add or multiply two numbers, if we know what type they are?

At first it might seem that we can. If we add two numbers that are each divisible by 3, the sum is again a multiple of 3; in other words, the sum of two threvens is threven. Likewise, if we add a number divisible by 3 to one that's not, the sum will *not* be a multiple of 3; the sum of a threven and a throdd is throdd. So far so good.

And now we hit the wall. If we add two throdd numbers, we can't tell which the sum will be: for example, the sum of 2 and 4—both throdd—is 6, which is threven; but the sum of 2 and 5—again both throdd—is 7, which is throdd.

Nor is this problem due to a particular badness of the number 3: in general, for any number n greater than 2 itself, the sum of two numbers not divisible by n may or may not be divisible by n; only for $n = 2$ is the outcome determined.

Does this mean there's nothing to be done, that the sort of information conveyed in Table 16-1 only makes sense when we talk about divisibility by 2 and not by other numbers? No, but it does mean that we need something more refined than the notion of throdd when we discuss nondivisibility by 3. When a number is odd, its remainder after division by 2 is necessarily equal to 1, but when a number is throdd, its remainder after division by 3 is either 1 or 2. If we want to determine the character of the sum of two throdd numbers, *we have to keep track of this remainder.*

This suggests defining three classes of numbers, instead of the two classes—threvens and throdds—we had before. The first class is the threvens—those numbers divisible by 3. The second class, which makes up half of the throdds, and which we will now call 1-throdds, consists of the numbers which leave a remainder of 1 after division by 3. For example, the throdds 4, 7, and 301 belong to the class of 1-throdds, but the throdds 5, 8 and 602 don't. The third class, again half of the throdds, which we will now call 2-throdds, consists of the numbers which leave a remainder of 2 after division by 3. Any whole number is either threven, 1-throdd, or 2-throdd, as these are the possible remainders after division by 3.

Now, you can check that the sum of a threven with a 1-throdd is a 1-throdd, and the sum of a threven with a 2-throdd is a 2-throdd. Similarly, the sum of a 1-throdd with a 2-throdd is a threven. Indeed, any 1-throdd has the form $3a + 1$ (meaning it is equal to $3a + 1$ for some whole number a) and any 2-throdd has the form $3b + 2$. But the sum is

$$(3a + 1) + (3b + 2) = 3a + 3b + 3 = 3(a + b + 1),$$

which is threven! Table 16-2 gives the complete addition table for the threvens, 1-throdds, and 2-throdds.

Now things look pretty much the same as our table for addition of evens and odds. We can also try to multiply our classes, and it turns out that multiplication is well defined. For example, the product of two 2-throdds is a 1-throdd. Why? Write the two 2-throdds to be multiplied as $3a + 2$ and $3b + 2$. Then the product

$$(3a + 2) \times (3b + 2) = 9ab + 6a + 6b + 4 = 3(3ab + 2a + 2b + 1) + 1$$

is clearly a 1-throdd: it's 3 times a number, plus 1. Table 16-3 is the complete multiplication table.

TABLE 16-2 Rules for Addition of Throdd and Threven Numbers

+	*threven*	*1-throdd*	*2-throdd*
threven	threven	1-throdd	2-throdd
1-throdd	1-throdd	2-throdd	threven
2-throdd	2-throdd	threven	1-throdd

TABLE 16-3 Multiplication of Throdd and Threven Numbers

×	*threven*	*1-throdd*	*2-throdd*
threven	threven	threven	threven
1-throdd	threven	1-throdd	2-throdd
2-throdd	threven	2-throdd	1-throdd

Exercise 16.2.1 Consider the numbers 4, 11 and 15.

1. Say whether each of these is threven, 1-throdd or 2-throdd
2. For each pair of the three numbers, say whether the sum of the two is threven, 1-throdd or 2-throdd. Does this agree with Table 16-2 above?
3. For each pair of the three numbers, say whether the product of the two is threven, 1-throdd or 2-throdd. Does this agree with Table 16-3 above?

16.3 Congruences

What is the moral of this story? What have we learned, and can we apply these lessons to issues of divisibility by numbers n other than 2 and 3? Actually, there's a number of things to be gleaned from this example, and we'll discuss these points in turn.

The first and plainest lesson is that if we break up all numbers into the class of those divisible by 3 and those not, we can't predict the class of a sum or product from the classes of the numbers being added or multiplied. But if we break the class of throdd numbers up further into the classes of those that leave a remainder 1 on division by 3 and those that leave a remainder of 2, we can. What does this tell us about divisibility by numbers other than 3? Well, to see how this would go if we replace 3 by 5, let's write out all the counting numbers in an array of five columns, like this:

$$\begin{array}{ccccc} 0 & 1 & 2 & 3 & 4 \\ 5 & 6 & 7 & 8 & 9 \\ 10 & 11 & 12 & 13 & 14 \\ 15 & 16 & 17 & 18 & 19 \\ 20 & 21 & 22 & 23 & 24 \\ 25 & 26 & 27 & \ldots & \end{array}$$

The left-hand column (which we'll call the 0 column, since it's the column that contains 0) consists of all the numbers divisible by 5; the next column (the 1 column, since it's the column that contains 1) contains all the numbers that leave a remainder of 1 after division by 5, and so on. We can extend this array backward into negative numbers as well:

	...	−13	−12	−11
−10	−9	−8	−7	−6
−5	−4	−3	−2	−1
0	1	2	3	4
5	6	7	8	9
10	11	12	13	14
15	16	17	18	19
20	21	22	23	24
25	26	27	...	

It may be a little dicey to talk about remainders when we're dealing with negative numbers, but there are other ways to characterize the columns. For example, we can say that the middle column consists of numbers "of the form $5a + 2$"—in other words, 2 more than a multiple of 5.

Having thus broken up all the whole numbers into five columns/classes, the key fact is that—very much analogously to what we saw when we broke up all whole numbers into threvens, 1-throdds and 2-throdds—*the column in which a sum appears depends only on the column in which the numbers being added appear.* Let's check this: for example, take a pair of numbers from the 2 column and the 4 column—say 2 and 9. Their sum, 11, is in the 1 column. What would happen if we chose a different pair of numbers from the 2 column and the 4 column? Try it: for example, we could take 7 and 14; they add up to 21, which is again in the 1 column. Or 12 and −6; their sum is 6, again in the 1 column. The fact is, *the sum of a number in the 2 column and a number in the 4 column will always lie in the 1 column.*

The analogous statement is true for multiplication: *the column in which a product appears depends only on the column in which the numbers being multiplied appear.* For example, if we multiply 2 and 9 we get 18, which is in the 3 column; if we multiply our other pairs, we find that

$$7 \times 14 = 98 \quad \text{and} \quad 12 \times -6 = -72$$

and both 98 and −72 appear in the 3 column as well (or would, if we extended the array). In other words, *the product of a number in the 2 column and a number in the 4 column will always lie in the 3 column.*

Before we go on, though, we need to address the issue of language: we need something to call the numbers in each of these classes/columns besides the equivalent of "2-throdd," or "in the 2 column." Instead, we will say a number that appears in the k column is *congruent to* k (mod 5). Other ways to say this are that it is "of the form $5a + k$", that is, equal to $5a + k$ for some whole number a; or that it leaves a remainder of k on division by 5. In fact, we'll extend this language: we'll say that any two numbers are congruent to each other (mod 5) if they appear in the same column of this array, that is, if they have the same remainder under division by 5. Another way to say this is that *two whole numbers a and b are congruent (mod 5) if their difference $a - b$ is divisible by* 5. We write this as $a \equiv b \pmod 5$.

(If you were extremely fastidious about notation, you might object here that we're using the symbol "≡" in two different contexts: we use it both to signify an equality between two numbers in the number system arithmetic mod 5, and to signify a relation between two ordinary whole numbers. But the two uses are at least consistent with each other, so no confusion is likely to arise—"no harm, no foul.")

Exercise 16.3.1 We saw above that the sum of a number congruent to 2 (mod 5) and one congruent to 4 (mod 5) is congruent to 1 (mod 5), and that their product is congruent to 3 (mod 5). Figure out the analogous statement for a pair of numbers

1. congruent to 2 and 3 (mod 5); and
2. congruent to 3 and 4 (mod 5).

It's now pretty clear how to extend this to an arbitrary modulus. First of all, we define the notion of congruence (mod n): we say that *two numbers a and b are congruent (mod n) if their difference $a - b$ is divisible by n*; that is, if they leave the same remainder under division by n. Note that every number is congruent (mod n) to one and only one of the numbers $0, 1, 2, \ldots, n-1$. Graphically, what we're doing is writing out all the whole numbers in an array of n columns:

$$\begin{array}{ccccccc} -n & -n+1 & -n+2 & \cdots & \cdots & -2 & -1 \\ 0 & 1 & 2 & \cdots & \cdots & n-2 & n-1 \\ n & n+1 & n+2 & \cdots & \cdots & 2n-2 & 2n-1 \\ 2n & 2n+1 & \cdots & & \cdots & 3n-2 & 3n-1 \end{array}$$

and saying that two numbers a and b are congruent (mod n) if they lie in the same column. In this case we write $a \equiv b \pmod{n}$.

Next, we make a key observation: that this way of breaking up all whole numbers into congruence classes (mod n) in fact does exactly what we set out to do. That is, it generalizes the notion of odd and even numbers, in the sense that

> The congruence class (mod n) of a sum or product is determined by the congruence classes (mod n) of the numbers added or multiplied.

Now this, so far, is just an observation. How would we convert it into a mathematical proof? There are infinitely many cases to check, so we can't do them individually. We have to devise a notation for an arbitrary member of the classes being added or multiplied (mod n). To do this, note that the numbers in the k column are those that can be written in the form $k + an$, for some number a. Those in the l column have the form $l + bn$, for a whole number b. If we calculate the sum, and rearrange the parentheses, we get:

$$(k + an) + (l + bn) = (k + l) + (a + b)n$$

which lies in the $k + l$ column. Similarly, if we calculate their product, and use the distributive law, we find:

$$\begin{aligned}(k + an) \times (l + bn) &= kl + kbn + aln + abn^2 \\ &= kl + (kb + al + abn)n.\end{aligned}$$

This is kl plus a multiple of n, and so lies in the kl column of the table. Hence, knowing the initial columns of the operation determines the final one.

Exercise 16.3.2 Take a pair of numbers congruent to 4 and 5 (mod 7). What is their sum congruent to (mod 7)? Try other pairs of numbers congruent to 4 and 5 (mod 7) and see if you get the same answer.

The next lesson to be learned from our calculations with threvens and throdds is more subtle. Take a look again at the tables we made of the rules for addition and multiplication of odd and even numbers, and compare these to the tables for arithmetic mod 2: if we substitute 0 and 1 for even and odd in the former, we get exactly the latter! Likewise, examine the tables 16-2 and 16-3 for combining threvens and throdds: if we replace the threvens by 0, the 1-throdds by 1, and the 2-throdds by 2, we get the tables for arithmetic mod 3 given in the last section. In general, the addition or multiplication table for arithmetic (mod n) gives us the rule for finding the congruence class (mod n) of a sum or product: writing

$$a + b \equiv c \pmod{n}$$

in the last chapter is equivalent in the language of this chapter to the statement that

> "the sum of a number congruent to a (mod n) and a number congruent to b (mod n) is always congruent to c (mod n)",

and similarly for multiplication.

The point is that in arithmetic mod n, we can either think of adding or multiplying remainders (mod n), which lie between 0 and $n - 1$, as we did in the previous section. Or, we can think of adding or multiplying classes of numbers with the same remainder after division by n, as we did in this section. Both points of view are equally interesting, although the latter is more flexible.

16.4 Calculating (mod n)

In this chapter, we've seen another viewpoint on arithmetic (mod n): we've learned to view the "numbers" in the number system arithmetic (mod n) not just as the symbols $0, 1, \ldots, n - 1$ (as we did in the last chapter) but as different classes of whole numbers, generalizing the notion of odd and even. The rules for adding, subtracting and multiplying, though, are unchanged: if we want to carry out any of these operations on two numbers a and b (mod n), we simply carry out those operations as we would in the ordinary whole number system, and then take remainders after division by n.

Well, not quite. It's true that we can in theory carry out any of these operations just by performing the corresponding operations on whole numbers, but there may also be easier ways of doing business, especially when the numbers involved become large. In the last section of this chapter, we'll show you a few of these tricks.

Let's start with an example involving just ordinary arithmetic. Suppose you have a calculator with a 12-digit display, and you're asked to evaluate the expression

$$\frac{254{,}191{,}101 \times 289{,}084}{437}.$$

Now, there are different ways to evaluate such an expression, depending on the order in which you carry out the operations. You could, for example,

- multiply 254,191,101 by 289,084 and then divide the result by 437; or you could
- divide 254,191,101 by 437, and then multiply the result by 289,084.

If you had infinite computational capacity, it wouldn't matter which way you chose to do the problem. But you don't, and it does. If you try to do the calculation the first way, when you try to multiply out the two factors you'll exceed the capacity of your calculator, and it'll give you an error message. (Or worse, it'll just round off and use exponential notation, without telling you that the answer is no longer 100% accurate.) If you do it the second way, though, there's no problem; the numbers stay within the capacity of the calculator.

Doing modular arithmetic is in some ways like this: there are different ways of carrying out a given calculation, and, while they are all theoretically equivalent, it may make a big difference in the degree of difficulty which you choose. To finish this chapter, we'll show you some examples of this.

We'll begin with a simple multiplication: find the product 224×376 in arithmetic mod 17. There are actually two ways to go about even as basic a calculation as this. We can multiply these two numbers as ordinary numbers:

$$224 \times 376 = 84{,}224$$

and then divide the result by 17 to find the remainder; that is, we would write

$$84{,}224 = 4{,}954 \times 17 + 6.$$

We arrive at the conclusion that

$$224 \times 376 \equiv 6 \pmod{17}.$$

But there's another way: we could *start* by taking remainders mod 17, and *then* multiply. If we do this, we find first that

$$224 \equiv 3 \pmod{17}$$

and

$$376 \equiv 2 \pmod{17}$$

so that

$$224 \times 376 \equiv 3 \times 2 \equiv 6 \pmod{17}.$$

This may not seem like a tremendous simplification—both ways of solving the problem can be carried out in a few seconds on a calculator—but imagine if we were dealing with numbers much larger than 224 and 376.

There are some other things we should be on the lookout for. For example, suppose we are working in arithmetic mod 633, and we want to calculate the product $439 \times 632 \pmod{633}$. One way to do it would be to multiply these two numbers as ordinary numbers:

$$439 \times 632 = 277{,}448$$

and then divide the result by 633 to find the remainder: that is, we would write

$$277{,}448 = 438 \times 633 + 194.$$

The result, then, is that

$$439 \times 632 \equiv 194 \pmod{633}.$$

But this is a lot of calculation, and even if this one can be done fairly quickly with a calculator, it's not hard to imagine similar problems with numbers too large to be handled by most calculators.

Fortunately, there's an easier way to do this particular problem. We observe that, mod 633, the number 632 is in the same class as -1: that is,

$$632 \equiv -1 \pmod{633}$$

so that

$$439 \times 632 \equiv 439 \times -1 \equiv -439 \quad \text{mod } 633.$$

Finally, to find what -439 is congruent to mod 633, we just add 633: $-439+633 = 194$, so $-439 \equiv 194 \pmod{633}$; and once again we see that the answer is 194.

(In fact, we've already seen this idea before, in Exercise 15.4.3—go back and take a look.)

Here's another problem of the same sort: find the square of 629 (mod 633). Again, we could multiply 629 by itself, divide the result by 633 and take the remainder to find the answer. But what's the point, when we have a faster method? All we really have to do is notice that $629 \equiv -4 \pmod{633}$, so that

$$629^2 \equiv (-4)^2 \equiv 16 \pmod{633}$$

and we see the answer is 16.

One more example, and then we'll let you try some on your own. Suppose we want to find the sixth power 4^6 of 4 (mod 17). The naive way to do this would be to multiply 4 by itself six times, to find that

$$4^6 = 4 \times 4 \times 4 \times 4 \times 4 \times 4 = 4{,}096.$$

We then divide by 17 and find that the remainder is 16, so that $4^6 \equiv 16 \pmod{17}$.

But there's another way, which avoids a lot of calculation: we can observe first that

$$4 \times 4 = 16 \equiv -1 \pmod{17}.$$

If we group the six factors of 4 in 4^6 into three groups of two—that is, we write

$$\begin{aligned} 4^6 &= 4 \times 4 \times 4 \times 4 \times 4 \times 4 \\ &= (4 \times 4) \times (4 \times 4) \times (4 \times 4) \\ &= (4^2)^3 \end{aligned}$$

—we see that

$$4^6 \equiv (-1)^3 \equiv -1 \pmod{17}$$

so that the answer is -1, or 16.

Now it's your turn:

Exercise 16.4.1 Carry out the following calculations in modular arithmetic:

1. $639 \times 437 \pmod 7$
2. $507 \times 237 \pmod{509}$
3. $367^2 \pmod{369}$
4. $7^6 \pmod{51}$

Exercise 16.4.2 Again, evaluate the following in modular arithmetic:

1. $432{,}903 + 1{,}463{,}974 \pmod{100}$
2. $105 \times 237 \pmod 7$
3. $4502^2 \pmod{4507}$
4. $76 \times 77 \times 78 \pmod{79}$

Chapter 17

Division

We have so far seen how to add, subtract and multiply numbers in modular arithmetic. In the following sections, we will add to these three more operations: we'll learn how to divide, how to take powers of a number, and how to take roots. Be prepared, though: these operations will involve a good deal more thought than the ones we've learned so far.

17.1 We Are Cleared for Takeoff

Division is often lumped together with addition, subtraction and multiplication as one of the four basic operations in arithmetic. (In *Alice in Wonderland*, The Mock Turtle describes the curriculum at his school as "Reeling and Writhing, of course, to begin with, . . . and then the different branches of Arithmetic—Ambition, Distraction, Uglification, and Derision.") In modular arithmetic, however, it is substantially more subtle than these, for two reasons:

- We can add, subtract and multiply any two numbers in any modular arithmetic system. The question of when we can divide, however, is far more delicate: in some modular number systems we can divide by any number other than 0; in others we can't.
- The operations of addition, subtraction and multiplication in arithmetic mod n can all be performed by carrying out the corresponding operations in ordinary whole-number arithmetic and then converting the answer to a number mod n. This is not the case with division: even when a quotient a/b exists, we may have to work a good bit harder to calculate it.

We will (as usual) approach this gingerly: we'll first say what we mean by the quotient of one number by another in any number system; then we'll investigate in a few examples when we can do this in arithmetic mod n; and finally we'll arrive at some general rules, both for *when* we can divide and *how* to actually do it.

17.2 What Are Quotients, and When Do They Exist?

The first part—saying what we mean by the ratio a/b of two numbers in any number system—is straightforward. We adopt the same definition as with subtraction: the quotient a/b of one number by another in any number system is the number c you multiply b by to get a—if one exists, and if there's no ambiguity. Note that we can always carry out division (by numbers other than 0) in fractions and decimals, but we may or may not be able to in counting numbers or whole numbers. We will also see number systems soon where there may be more than one number c in the system such that $c \times b = a$, and we'll talk about this.

As for the second part, let's start with our first example, arithmetic mod 5. Can we divide in arithmetic mod 5? To answer this, look again at the multiplication table for arithmetic mod 5, which we'll reproduce here.

TABLE 17-1 The Multiplication Table for Arithmetic Mod 5

×	*0*	*1*	*2*	*3*	*4*
0	0	0	0	0	0
1	0	1	2	3	4
2	0	2	4	1	3
3	0	3	1	4	2
4	0	4	3	2	1

For example, let's evaluate the fraction $2/3$ in arithmetic mod 5. What this means is, find the number in arithmetic mod 5 that, when you multiply it by 3, you get 2. We can do this by looking at the table: if we look at the "3" row, we see a 2 in the "4" column. This means that $4 \times 3 \equiv 2 \pmod{5}$, so that 4 is the answer we seek—in other words,

$$\frac{2}{3} \equiv 4 \pmod{5}.$$

Let's do another one: say we want to find the fraction $1/2$ in arithmetic mod 5. As before, we look at the "2" row of our table, and look for a 1, which we find in the "3" column; that means $3 \times 2 \equiv 1 \pmod{5}$, so that

$$\frac{1}{2} \equiv 3 \pmod{5}.$$

For one more example, let's try a division in arithmetic mod 7. To do this, we'll need the multiplication table for this number system, which is given in Table 17-2 on the following page.

With this, we should be able to find, for example, the fraction $6/5$—that is, the number you multiply 5 by to get 6 in arithmetic mod 7—if it exists. And it does: we look in the "5" row, we see a 6 in the "4" column, and deduce that

$$\frac{6}{5} \equiv 4 \pmod{7}.$$

TABLE 17-2 The Multiplication Table for Arithmetic Mod 7

×	0	1	2	3	4	5	6
0	0	0	0	0	0	0	0
1	0	1	2	3	4	5	6
2	0	2	4	6	1	3	5
3	0	3	6	2	5	1	4
4	0	4	1	5	2	6	3
5	0	5	3	1	6	4	2
6	0	6	5	4	3	2	1

Now you do some:

Exercise 17.2.1 Using the tables above, evaluate

1. The quotient $4/3$ in arithmetic mod 5;
2. The quotient $4/3$ in arithmetic mod 7; and
3. The quotient $3/4$ in arithmetic mod 7.

Exercise 17.2.2 Consider the expression

$$\frac{2 \times 3}{5} \pmod{7}.$$

Evaluate this four ways:

1. Multiply 2 by 3 and divide the result by 5;
2. Divide 2 by 5 and multiply the result by 3;
3. Divide 3 by 5 and multiply the result by 2;
4. Find the reciprocal $1/5$ of 5 in arithmetic mod 7, and multiply this by 2 and then by 3.

Let's stop here and ask a qualitative question: can we *always* carry out a division a/b in arithmetic mod 7, as long as the denominator b isn't 0?

To answer this, look again at the tables. The key fact we see here is that *every row of this table except the "0" row contains every number in the number system exactly once*. What this means is that the answer to our question is "yes:" in arithmetic mod 7 we can always divide by any nonzero number.

The next question is, why is this? Why does every number appear in every row? The answer is, since there are 6 nonzero numbers in arithmetic mod 7 and (not coincidentally) 6 columns in our table other than the "0" column, a row can fail to contain every number only if some number appears twice or more. But what would it mean if the same number appeared twice in a row—say, for example, that in the "a" row, the same number appeared in both the "b" column and the "c" column? It would mean that

$$b \times a \equiv c \times a \pmod{7}$$

that is, by the distributive law,

$$(b - c) \times a \quad (\text{mod } 7).$$

In other words, we'd have two numbers, $b - c$ and a, both between 1 and 6, whose product was divisible by 7. *But that can't happen because 7 is prime*: if the product of two numbers is divisible by 7, one or the other must be.

The same logic, in fact, applies in arithmetic mod p for any prime number p: every nonzero number must appear in every row other than the "0" row in the multiplication table for arithmetic mod p, and we deduce the important fact that

> If p is a prime number, then in arithmetic mod p we can divide by any number except 0.

Contrast this with, for example, the situation in arithmetic mod 6. Here's the relevant multiplication table:

TABLE 17-3 The Multiplication Table for Arithmetic Mod 6

×	0	1	2	3	4	5
0	0	0	0	0	0	0
1	0	1	2	3	4	5
2	0	2	4	0	2	4
3	0	3	0	3	0	3
4	0	4	2	0	4	2
5	0	5	4	3	2	1

What we see is that not every number appears in each row, and as a result we can't always divide: as you can see, the ratios $2/3$, $3/4$ and $5/2$ don't exist. But we also notice that there are two rows that do contain every nonzero number in arithmetic mod 6: the "1" row and the "5" row. Why? *Because these are the two numbers between 1 and 5 that are relatively prime to 6.* (If you're a little rusty on the concept of "relatively prime," now might be a good time to review the first few sections of Chapter 13.) The point is, if we look in the multiplication table for arithmetic mod n at the "a" row, where a is any number relatively prime to n, the same logic applies: the same number can't appear in two different columns; therefore every number must appear once; therefore we can always divide by a. We can thus strengthen the statement above:

> In arithmetic mod n, we can divide by any number a relatively prime to n.

17.3 How to Divide

When we said, in the last section, that "we can always divide" by any nonzero number in arithmetic mod p, what we meant is really that "the ratio a/b always exists (as long as $b \neq 0$)"—in other words, there will always be some number

c in arithmetic mod p that you can multiply b by to get a. But it's one thing to say division is possible; it's quite another to actually find the quotient a/b of two numbers. In this section, we'll see how you can actually carry out such a calculation.

To see how this goes, let's take a sample calculation and see how to do it in different ways. For our sample problem, let's try finding the reciprocal $1/7$ in arithmetic mod 11. There are a number of ways we might go about it.

- We could write out the entire multiplication table for arithmetic mod 11, and then consult the table as we've done in similar situations before. This is somewhat inefficient when we're dealing with a relatively small modulus like 11, and completely impractical if we had to do it for a larger modulus: writing out the multiplication table for arithmetic mod n involves something like n^2 steps. The point is, we don't need the entire multiplication table in order to divide by 7; we only need the "7" row.
- Here is a more efficient method. Write out the "7" row of the multiplication table for arithmetic mod 11 until you get a 1. In other words, work out all the multiples of 7 (mod 11) until you find the one you want:

$$1 \times 7 \equiv 7 \pmod{11}$$
$$2 \times 7 \equiv 3 \pmod{11}$$
$$3 \times 7 \equiv 10 \pmod{11}$$
$$4 \times 7 \equiv 6 \pmod{11}$$
$$5 \times 7 \equiv 2 \pmod{11}$$
$$6 \times 7 \equiv 9 \pmod{11}$$
$$7 \times 7 \equiv 5 \pmod{11}$$
$$8 \times 7 \equiv 1 \pmod{11}$$

  Bingo! $8 \times 7 \equiv 1 \pmod{11}$, so we have

$$\frac{1}{7} \equiv 8 \pmod{11}.$$

  Note that if we're working with a large modulus n, this process—taking multiples of the denominator until you find the one you want—is guaranteed to work in n or fewer steps; on the average, it'll take roughly $n/2$ steps. In other words, it's a reasonable approach if the modulus is in the single digits or teens; and more than slightly masochistic for larger moduli.
- Another reasonably straightforward way: find a number congruent to 1 (mod 11) that is divisible by 7. What we're trying to do here, after all, is to find a multiple of 7 that is congruent to 1 (mod 11). Last time we did this by listing all the numbers divisible by 7 until we found one congruent to 1 (mod 11); this time we'll turn that around and make a list of all the numbers congruent to 1 (mod 11) until we find one that is divisible by 7:

$$1 \times 11 + 1 = 12$$
$$2 \times 11 + 1 = 23$$
$$3 \times 11 + 1 = 34$$
$$4 \times 11 + 1 = 45$$
$$5 \times 11 + 1 = 56 = 8 \times 7$$

Bingo! Again, we see that $8 \times 7 \equiv 1 \pmod{11}$, so that $\frac{1}{7} \equiv 8 \pmod{11}$. As with the last method, in arithmetic mod n this will always work after at most n steps, and on average in $n/2$ steps.

- The smart way. After all, what we're trying to do in both of the last two approaches is simply to find a multiple of 7 that is congruent to 1 (mod 11). To put it another way, we want to solve the equation

$$7x = 11y + 1.$$

But *we already know how to do this!* Solving equations like this—expressing 1 as a combination of 7 and 11, in other words—is exactly what we learned how to do in Sections 9.3 and 9.4. (If you're a little rusty on that material—it was a while ago, after all—it would be a good idea to take a moment out and review it now.) To carry this out, we just run the Euclidean algorithm to find the greatest common divisor of 7 and 11:

$$\begin{aligned} 11 &= 1 \times 7 + 4 \\ 7 &= 1 \times 4 + 3 \\ 4 &= 1 \times 3 + 1. \end{aligned}$$

We then run it backwards to express 1 as a combination of 7 and 11:

$$\begin{aligned} 1 &= 4 - 3 \\ &= 4 - (7 - 4) \\ &= 2 \times 4 - 7 \\ &= 2 \times (11 - 7) - 7 \\ &= 2 \times 11 - 3 \times 7. \end{aligned}$$

(Of course, we knew going in that gcd(7, 11) would be 1, since 11 is prime and 7 is not divisible by 11. But we had to run the Euclidean algorithm anyway, in order to carry out the second half of the process.) Now, what this last equation is telling us that 1 differs from the product -3×7 by a multiple of 11; or, in other words, that -3 times 7 is congruent to 1 (mod 11). This is the answer to our question: -3 is the number you multiply 7 by to get 1 in arithmetic mod 11, so that

$$\frac{1}{7} \equiv -3 \equiv 8 \pmod{11}.$$

As you can see, this last method is not any easier in the particular problem we're trying to solve here. But when the numbers get large, this is the only practical way to carry out a division.

Let's look at one more example of this method, in a case where none of the other methods would be feasible.

EXAMPLE 17.3.1 Find the quotient $\frac{7}{216}$ in arithmetic (mod 691).

SOLUTION We are trying to find a multiple of 216 that is congruent to 7 (mod 691)—in other words, we want to solve the equation

$$216x = 691y + 7.$$

As we saw in Sections 9.3 and 9.4, the way to do this is to first solve the equation

$$216x = 691y + 1.$$

To do this, we run the Euclidean algorithm to compute the gcd of 691 and 216:

$$691 = 3 \times 216 + 43$$
$$216 = 5 \times 43 + 1$$

and then run it backwards to express 1 as a combination of 216 and 691:

$$\begin{aligned} 1 &= 216 - 5 \times 43 \\ &= 216 - 5(691 - 3 \times 216) \\ &= 16 \times 216 - 5 \times 691. \end{aligned}$$

In other words, we see that $16 \times 216 \equiv 1 \pmod{691}$, so that

$$\frac{1}{216} \equiv 16 \pmod{691}.$$

Now we just multiply by 7 to find that

$$\frac{7}{216} \equiv 7 \times 16 \equiv 112 \pmod{691}.$$

■

Notice one thing about this method: it works by virtue of the fact that the denominator 216 is relatively prime to the modulus 691, which is *exactly the condition that we are able to divide at all.* In other words, if we tried to use this method to find a quotient that didn't exist, the Euclidean algorithm would simply tell us that we couldn't express 1 as a combination of the denominator and modulus.

We'll see a couple other ways to do division in modular arithmetic, one in the following section and one further on in Section 18.2; but for now it's time for you to get some practice.

Exercise 17.3.2 Carry out the following divisions in modular arithmetic:

1. Find $3/8 \pmod{13}$;
2. Find $5/17 \pmod{60}$; and
3. Find $23/61 \pmod{127}$.

Exercise 17.3.3 Rebecca is running a repetitive program on her computer. One cycle of this program takes exactly 43 seconds to run. Rebecca starts the program when the second hand of her watch is exactly at the top. Some time later (less than half an hour) she notices that just as the computer completes a cycle the second hand of her watch is exactly 5 seconds past the top—in other words, the number of seconds elapsed since the program started is congruent to 5 (mod 60). How many cycles has the program completed?

17.4 Reciprocals

An important special case of division in general is the finding of *reciprocals*—that is, for each number a in arithmetic mod n, finding the number $c = 1/a$ you multiply a by to get 1. In fact, once we know the reciprocals of all the numbers in arithmetic mod n, we can carry out any division a/b just by multiplying the reciprocal $1/b$ by a.

Thus, if we're planning on doing a lot of arithmetic in a particular modular arithmetic system, it may make sense to make a table of reciprocals in that number system. To illustrate this, and because it'll be a good chance to show off some of the tricks we've learned, we'll do this for arithmetic mod 13.

To do this, we'll write out the numbers $1, 2, \ldots, 12$ in arithmetic mod 13, and below them we'll write their reciprocals. To begin with, there are two obvious ones: the reciprocal of 1 is 1 in any modulus; and likewise the reciprocal of -1 is always -1. We start by entering this information in our table:

1	2	3	4	5	6	7	8	9	10	11	12
1											12

Now, there are some numbers whose reciprocals we can find immediately. For example, to find the fraction $1/2$, we just have to find a multiple of 2—an even number, in other words—that's congruent to 1 (mod 13). No sweat: $14 = 7 \times 2$ does the trick, so that

$$\frac{1}{2} \equiv 7 \pmod{13}.$$

Of course, that's not all we learn from this: if $1/2 \equiv 7 \pmod{13}$, then likewise we must have

$$\frac{1}{7} \equiv 2 \pmod{13}.$$

We enter this information in our table as well:

1	2	3	4	5	6	7	8	9	10	11	12
1	7					2					12

Now, we can do more with this: after all, if $1/2 \equiv 7 \pmod{13}$, since $2 \times 2 = 4$, it follows that

$$\frac{1}{4} \equiv \frac{1}{2} \times \frac{1}{2} \equiv 7 \times 7 \equiv 10 \pmod{13}.$$

and likewise

$$\frac{1}{10} \equiv 4 \pmod{13}.$$

Let's enter this into our table as well:

1	2	3	4	5	6	7	8	9	10	11	12
1	7		10			2			4		12

Likewise, having found the reciprocal $1/2 \equiv 7 \pmod{13}$ of 2, we now know the reciprocal of -2 as well:

$$\frac{1}{-2} \equiv -7 \equiv 6 \pmod{13}$$

so that

$$\frac{1}{11} \equiv 6 \quad \text{and} \quad \frac{1}{6} \equiv 11 \pmod{13}.$$

Our table is now nearly complete:

1	2	3	4	5	6	7	8	9	10	11	12
1	7		10		11	2			4	6	12

There are many ways to polish this off. Another way to add entries to the table, for example, is instead of looking for reciprocals of specific numbers, we can just look for pairs of numbers that multiply to 1 in arithmetic (mod 13). The way to do this is to list the whole numbers that are congruent to 1 (mod 13), and then factor them: for example,

$$2 \times 13 + 1 = 27 = 3 \times 9$$

and

$$3 \times 13 + 1 = 40 = 5 \times 8$$

which tells us that

$$\frac{1}{3} \equiv 9 \qquad \text{and} \qquad \frac{1}{9} \equiv 3 \pmod{13}$$

and

$$\frac{1}{5} \equiv 8 \qquad \text{and} \qquad \frac{1}{8} \equiv 5 \pmod{13}$$

respectively. This completes our table:

TABLE 17-4 Reciprocals Mod 13

1	2	3	4	5	6	7	8	9	10	11	12
1	7	9	10	8	11	2	5	3	4	6	12

Exercise 17.4.1 Use the table to evaluate the fractions 4/9, 8/7 and 5/6 in arithmetic (mod 13)

Exercise 17.4.2 Write out a similar table of reciprocals for arithmetic mod 7 and mod 11. Are there any numbers that are their own reciprocals, other than 1 and -1? Is this a general pattern in arithmetic mod p, when p is a prime? What happens in arithmetic mod 8?

17.5 A Warning

We've seen in this chapter that, in arithmetic mod n, we can always divide by any number a relatively prime to n. It's important to note also that, conversely, *it does not make sense to talk about quotients a/b where the denominator is not relatively prime to the modulus n.*

This is the case for two reasons. One we've already seen: if the denominator b is not relatively prime to n, the quotient a/b simply may not exist, in the sense that there may not be any number c in arithmetic mod n such that

$$b \cdot c \equiv a \pmod{n}.$$

For example, in arithmetic mod 12, the ratio $1/2$ doesn't exist: there is no multiple of 2 congruent to 1 (mod 12).

But there's another, more insidious reason why dividing by numbers not relatively prime to the modulus is bad, and this is the real reason why we're stressing this warning: in cases where the fraction does exist, it will not be unique, and this ambiguity can lead to outright fallacy.

Here's an example: consider the fraction $6/3$ in arithmetic mod 12. You might look at this and say, "What's the big deal? $6/3$ is 2 in any number system, because $2\times 3 = 6$ in any number system. What does it matter if the modulus isn't relatively prime to the denominator?" By the same token, it seems harmless enough to say that the fraction $6/2 \equiv 3 \pmod{12}$. But now we multiply these two fractions, and everything goes haywire: on the one hand, if $6/3 \equiv 2 \pmod{12}$ and $6/2 \equiv 3 \pmod{12}$, then surely

$$\frac{6}{3} \times \frac{6}{2} \equiv 2 \cdot 3 \equiv 6 \pmod{12}.$$

But on the other hand, it seems equally clear that

$$\frac{6}{3} \times \frac{6}{2} \equiv \frac{6 \times 6}{3 \times 2} \equiv \frac{0}{6} \equiv 0 \pmod{12}.$$

What's going wrong here? Basically, the problem is that there is more than one possible value for the fractions $6/3$ and $6/2$: for example, we said that $6/3 \equiv 2 \pmod{12}$ because $2\times 3 \equiv 6 \pmod{12}$. But it's equally true that $6\times 3 \equiv 6 \pmod{12}$ and $10 \times 3 \equiv 6 \pmod{12}$, so we could equally well assign the values 6 and 10 to the ratio $6/3 \pmod{12}$. Similarly, $9 \times 2 \equiv 6 \pmod{12}$, so we could just as well say $6/2 \equiv 9 \pmod{12}$. When we treat the expression "$6/3 \pmod{12}$" as if it were meaningful, in other words, we are in effect equating all the different possible values 2, 6 and 10 of the quotient, and so of course we get into trouble.

Just remember: dividing by numbers not relatively prime to the modulus is bad. Don't do it.

HELP ME FIGURE OUT THIS HOMEWORK PROBLEM, HOBBES. WHAT'S 3+8?
OK, ASSIGN THE ANSWER A VALUE OF "X". "X" ALWAYS MEANS MULTIPLY, SO TAKE THE NUMERATOR (THAT'S LATIN FOR "NUMBER EIGHTER") AND PUT THAT ON THE OTHER SIDE OF THE EQUATION.
©1988 Universal Press Syndicate
THAT LEAVES YOU WITH THREE ON THIS SIDE, SO WHAT TIMES THREE EQUALS EIGHT? THE ANSWER, OF COURSE, IS SIX.
GOSH, I MUST HAVE DONE ALL THE OTHERS WRONG.
THESE PROBLEMS SEEM AWFULLY ADVANCED FOR FIRST GRADE, IF YOU ASK ME.
1-5
WATTERSON

Chapter 18

Powers

There's a classic fable about powers ...

A long time ago, the ruler of a kingdom was introduced to the game of chess by a merchant. Pleased with the game, he summoned the merchant to his throne.

"O loyal subject," the king said, "we have decided to reward you for this game by giving you one gold coin for each of the 64 *squares of the chessboard."*

"O supreme monarch," the merchant replied, "I am not worthy to receive such a reward. However, if my august majesty desires to reward his insignificant servant, he might place on the first square of the chessboard one simple grain of rice; on the second, two grains; on the third, four; on the fourth, eight; and so on."

The ruler immediately agreed to this, and called forth his cook with the rice canister. One grain was placed on the first square, two on the second, and so forth, the king all the time thinking of the gold coins he was saving by this arrangement.

By the end of the first row, they were up to 128 *grains—roughly a tablespoonful. By the end of the second row, the amount had gone up to* 32,768 *grains, about a gallon in volume. After three rows, it was* 8,388,608*, or roughly a hogshead; after four, a boxcarful; and at the end of the sixth row the amount exceeded all the rice in the kingdom. Realizing the impossibility of continuing any further, the king called in his guards and had the merchant taken away and beheaded.*

There are two morals to this story:

- Don't back someone with supreme power into a corner; and
- The powers of 2 grow faster than you might think.

We will be primarily concerned here with the second of these morals (though the first is also a valuable life lesson).

In this chapter we're going to consider how powers of a number behave in modular arithmetic. Before we begin, though, be aware that the sort of questions we're going to ask are of necessity very different from what you may expect. After all, in ordinary arithmetic, the powers of any number bigger than 1 just keep growing larger and larger, and the main question we might ask is, "how fast?" In arithmetic mod n, by contrast, of necessity the powers of a given number don't keep growing: they just bounce around among the n numbers $\{0, 1, 2, \ldots, n-1\}$, and what we'll be looking for is patterns in the way in which these numbers are hit.

18.1 Preliminaries Redux

Before we begin our investigation in earnest, we want to review the two basic rules for combining powers of a number. We mentioned these before, in Section 12.1, but they bear repeating.

The first rule is for multiplying two powers of a number x. If we multiply the ath power of x by the bth power, the result is simply the product of x with itself $a+b$ times—that is, the $(a+b)^{\text{th}}$ power of x:

$$\underbrace{(x \times x \times \cdots \times x)}_{a} \times \underbrace{(x \times \cdots \times x)}_{b} = \underbrace{x \times x \times x \times \cdots \times x \times x}_{a+b}$$

or in other words

$$x^a \times x^b = x^{a+b}.$$

Next, suppose we take the ath power of a number x, and then raise that number to the bth power. We can write this out as the product of a string of x's

$$\underbrace{\underbrace{(x \times \cdots \times x)}_{a} \times \underbrace{(x \times \cdots \times x)}_{a} \times \ldots \underbrace{(x \times \cdots \times x)}_{a}}_{b} = \underbrace{(x \times x \times x \times \cdots \times x \times x)}_{ab}$$

or in other words

$$(x^a)^b = x^{ab}.$$

Note that to establish these rules we need only apply the second of the axioms for a number system, so that they hold true in any number system.

We should also say again that it's important to bear these formulas in mind even when we're not prompted to. Here's what we mean: when we see an expression like $(x^3)^4$, this prompts us to remember the second of the two formulas above; and once we think of it we know we can multiply this out to get x^{12}. But there'll be a lot of occasions in the next few chapters where we'll see an expression like x^{12}, and it'll be important to remember that this can also be rewritten as $(x^6)^2$, or $(x^4)^3$, or $(x^3)^4$, and so on—even though the expression x^{12} won't necessarily prompt us to think of the formula. In other words, bear in mind that these are, as we said in Section 12.1, two-way formulas.

18.2 Calculating Powers

We want to take a moment here and show how you can use the rules in the last section to calculate powers more efficiently.

The issue here is not whether we can evaluate a power of a given number, but how many steps it'll take us to do it. Suppose, for example, that someone asked you to find 3^{32}. Of course, you could start with 3, multiply by 3 to get 3^2, multiply by 3 again to get 3^3, and so on; after 31 operations, you'd have the answer.

But there's a better way. The formulas in the last section tell us in particular that, for any numbers x and a,

$$(x^a)^2 = x^{2a}.$$

In other words, if we know the ath power of a number x, we can square it to get the $(2a)^{\text{th}}$ power. In the present circumstance, we apply this with $x = 3$, and we write

$$\begin{aligned} 3^2 &= 9 \\ 3^4 &= (3^2)^2 = 9^2 = 81 \\ 3^8 &= (3^4)^2 = 81^2 = 6{,}561 \\ 3^{16} &= (3^8)^2 = 6{,}561^2 = 43{,}046{,}721 \\ 3^{32} &= (3^{16})^2 = 43{,}046{,}721^2 = 1{,}853{,}020{,}188{,}851{,}841, \end{aligned}$$

arriving at the answer in five steps.

Well, you could say, we were lucky that time: the exponent 32 happened to be a power of 2, so we could arrive at 3^{32} by successive squaring. But in fact a variant of this method works for any exponent a, because any number a can be written as a sum of various powers of 2. Suppose for example that we wanted to find 3^{27} instead of 3^{32}. We start by writing 27 as a sum of powers of 2:

$$27 = 16 + 8 + 2 + 1$$

then we consult the table above to find the values of 3^{16}, 3^8, 3^2 and 3^1, and we multiply them out:

$$\begin{aligned} 3^{27} &= 3^{16} \times 3^8 \times 3^2 \times 3^1 \\ &= 43{,}046{,}721 \times 6{,}561 \times 9 \times 3 \\ &= 7{,}625{,}597{,}484{,}987. \end{aligned}$$

This process takes an extra 3 steps—multiplying out the factors 3^{16}, 3^8, 3^2 and 3^1—but it's still a lot shorter than multiplying 3 by itself 26 times,

We'll go through the process one more time in general, in parallel with a concrete example. The example we'll choose will be in modular arithmetic, rather than the ordinary kind; remember that *this method of computing powers works in any number system*, because the rules it's based on do. It's particularly useful in modular arithmetic, since (as we'll see) in modular arithmetic we often want to deal with extremely large exponents.

Method for Calculating Powers

To find a high power x^a of a number—for example, say, 7^{42} (mod 11)—we do the following:

- ***Express the exponent a as a sum of different powers of 2.*** To do this, first find the largest power of 2 that is still less than a. Subtract that from a, and then find the largest power of 2 less than the remainder; subtract that and continue until nothing is left.

  In the example, the exponent $a = 42$. The largest power of 2 less than 42 is $2^5 = 32$, so that'll be the first term. Now $42 - 32 = 10$, and the largest power of 2 less than 10 is $2^3 = 8$; so that's second. Finally, what's left, $10 - 8 = 2$ is a power of 2, so we're done: we have

$$42 = 32 + 8 + 2.$$

- ***Make a table of powers of x by successive squaring.*** That is, start with x, square it to get x^2, square that to get x^4, square that to get x^8, and so on until the exponent is the largest power of 2 less than a.

  In our case, the base $x = 7$, so we make a table:

$$
\begin{aligned}
7^1 &\equiv 7 \\
7^2 &\equiv 49 \equiv 5 \pmod{11} \\
7^4 &\equiv (7^2)^2 \equiv 5^2 \equiv 25 \equiv 3 \pmod{11} \\
7^8 &\equiv (7^4)^2 \equiv 3^2 \equiv 9 \pmod{11} \\
7^{16} &\equiv (7^8)^2 \equiv 9^2 \equiv 81 \equiv 4 \pmod{11} \\
7^{32} &\equiv (7^{16})^2 \equiv 4^2 = 16 \equiv 5 \pmod{11}
\end{aligned}
$$

  We stop at 32 because that's the largest power of 2 less than the exponent 42.

- ***Multiply the powers of x in your table to get x^a.*** By our first calculation, we can express x^a as a product of powers of x whose exponents are powers of 2; by the second, we know what those particular powers are, and we can just multiply them out to get the answer.

  In our example, we know from Step 1 that

$$7^{42} = 7^{32} \times 7^8 \times 7^2$$

  and so, using the list we generated in Step 2,

$$
\begin{aligned}
7^{42} &\equiv 7^{32} \times 7^8 \times 7^2 \pmod{11} \\
&\equiv 5 \times 9 \times 5 \pmod{11} \\
&\equiv 5 \pmod{11}
\end{aligned}
$$

Now it's your turn:

Exercise 18.2.1 Use the method described here to find $11^{55} \pmod{19}$.

18.3 Laws for Powers

We now begin our investigation of laws for powers in modular arithmetic. As we said, powers in modular arithmetic necessarily behave very differently from powers in ordinary arithmetic: whereas in ordinary arithmetic the powers of any number bigger than 1 just keep getting bigger and bigger, the powers of a number x in arithmetic mod n have to keep bouncing around among the n numbers $\{0, 1, 2, 3, \ldots, n-1\}$. In particular, since there are only finitely many numbers in arithmetic mod n, we see that the powers of any number x can't all be different; *eventually they have to repeat*, a phenomenon that is completely unlike our experience of ordinary arithmetic. This raises all sorts of questions: Is there a pattern in the powers of a given number in arithmetic mod n? Can we predict when they'll start to repeat? We'll look into this now, and in fact will come up with a nearly complete answer to this question.

As usual, we'll start by doing a bunch of examples, to see what sort of behavior we might expect. One thing we want to stress: it's important that you do some of this on your own. Read through this section, but then before going on to the next section, where the answer is revealed, try working out some additional examples on your own and seeing the pattern for yourself. We've said this before—it's true of everything we do in this book—but here in particular we can't emphasize it too strongly: if you want to stay with us through the next three chapters, stop every now and then and work out some examples on your own to get a feel for how these numbers behave.

What examples should we do to start with? Well, one thing we noticed in the last section is that multiplication and division (and hence taking powers) in arithmetic mod n behaves very differently depending on whether n is a prime number or not. In our present context, we might note that in modular arithmetic a power of a number can be zero even when that number isn't: for example, in arithmetic mod 24, the number 6 is not zero, and its square $6^2 = 36 \equiv 12 \pmod{24}$ isn't, but $6^3 = 216 \equiv 0 \pmod{24}$ is. But *this can't happen if the modulus is prime.* For example, in arithmetic mod 37, to say that a power x^a of a number x is zero is to say that in ordinary arithmetic the number x^a is divisible by 37. But as we've seen, if 37 divides $x \times x \times \cdots \times x$ it must divide x—in other words, x must have been congruent to 0 (mod 37) to begin with.

Accordingly, since we're just starting out, *we're going to restrict ourselves, for the remainder of this chapter, to the case of arithmetic mod p, where p is a prime number*. We'll leave the case where the modulus is composite to a later chapter.

That said, let's start with a nice small modulus—say, 7. (We'll leave 5 for you to work out on your own.) We begin by making a list of the powers of 2 (mod 7):

$$2^1 \equiv 2 \pmod{7}$$
$$2^2 \equiv 4 \pmod{7}$$
$$2^3 \equiv 1 \pmod{7}$$
$$2^4 \equiv 2 \pmod{7}$$
$$2^5 \equiv 4 \pmod{7}$$
$$2^6 \equiv 1 \pmod{7}$$
$$2^7 \equiv 2 \pmod{7}$$

and so on. We can stop here (we could have stopped a few steps back, for that matter), because all that's going to happen is that the cycle 2, 4, 1 is going to repeat itself over and over forever. Can you say why?

Now let's do the same thing for powers of 3:

$$3^1 \equiv 3 \pmod{7}$$
$$3^2 \equiv 2 \pmod{7}$$
$$3^3 \equiv 6 \pmod{7}$$
$$3^4 \equiv 4 \pmod{7}$$
$$3^5 \equiv 5 \pmod{7}$$
$$3^6 \equiv 1 \pmod{7}$$
$$3^7 \equiv 3 \pmod{7}$$
$$3^8 \equiv 2 \pmod{7}$$
$$3^9 \equiv 6 \pmod{7}$$
$$3^{10} \equiv 4 \pmod{7}$$
$$3^{11} \equiv 5 \pmod{7}$$
$$3^{12} \equiv 1 \pmod{7}$$
$$3^{13} \equiv 3 \pmod{7}$$

Again, there's no need to continue: the cycle 3, 2, 6, 4, 5, 1 is just going to repeat itself.

It may be more efficient to present this data in a table:

	x	x^2	x^3	x^4	x^5	x^6	x^7	x^8	x^9	x^{10}	x^{11}	x^{12}	x^{12}
2	2	4	1	2	4	1	2	4	1	2	4	1	2
3	3	2	6	4	5	1	3	2	6	4	5	1	3

While we're at it, we'll complete this table by adding rows for 1, 4, 5 and 6, and give the result in Table 18-1

TABLE 18-1 Powers Mod 7

	x	x^2	x^3	x^4	x^5	x^6	x^7	x^8	x^9	x^{10}	x^{11}	x^{12}	x^{13}
1	1	1	1	1	1	1	1	1	1	1	1	1	1
2	2	4	1	2	4	1	2	4	1	2	4	1	2
3	3	2	6	4	5	1	3	2	6	4	5	1	3
4	4	2	1	4	2	1	4	2	1	4	2	1	4
5	5	4	6	2	3	1	5	4	6	2	3	1	5
6	6	1	6	1	6	1	6	1	6	1	6	1	6

There are already a number of fascinating patterns here. But before we go on and investigate, it's important that you do the following exercise:

Exercise 18.3.1 Make a similar table of the powers of the nonzero numbers mod 5, up to and including the 9th power of each number.

Now let's take a look at Table 18-1. We'll make a few observations, starting with the most obvious. We will express these all for arithmetic mod 7, but as you can see they apply to any flavor of modular arithmetic, as long as the modulus is prime. First of all,

- There have to be repetitions in each row, that is, in the list of powers of any given number. Why? Because there are only six nonzero numbers in arithmetic mod 7, and there are infinitely many powers to compute, so there have to be some repetitions.
- There has to be a 1 in each row; that is, some power of any given number has to be 1. We just said there have be repetitions among the powers of a given number x; that is, there must be two different exponents a and b, with $a < b$, such that

  $$x^a \equiv x^b \pmod{7}.$$

  But since $x^b = x^a \times x^{b-a}$, we can rewrite this as

  $$x^a \equiv x^a \times x^{b-a} \pmod{7}.$$

  Now, we can divide by x^a (remember that $x^a \not\equiv 0 \pmod{7}$, and that we can divide by any nonzero number modulo a prime), so this tells us that

  $$x^{b-a} \equiv 1 \pmod{7}.$$

  So we see that some power of any given number must be 1.

- Once we see a 1 in a row, the row starts repeating. Again, if

$$x^c \equiv 1 \pmod 7,$$

then necessarily

$$x^{c+1} \equiv x \pmod 7,$$

and

$$x^{c+2} \equiv x^2 \pmod 7,$$

and so on, so the row just starts over from the beginning.

This is the picture in general: the powers of a given number x will run through a group of different numbers until the first power x^c congruent to 1; from that point on, the powers just repeat themselves in cycles of length c. What's more, this is the case for powers mod p for any prime p.

What's the next question? There are many things we might want to ask, but there's one thing that stands out in Table 18-1: the "x^6" column is all 1s. Why should that be? (And is there an analogous statement about powers mod 5, based on your answer to Exercise 18.3.1?) The "x^{12}" column also consists entirely of 1s, for that matter; and the "x^7" and "x^{13}" columns are just repetitions of the "x" column, but these are all just consequences of the fact that the "x^6" column is all 1s, given the way each row goes in cycles. What this suggests is the question: *if the powers of a given number mod p repeat themselves in cycles, what can we say about the lengths of these cycles?*

According to Table 18-1, the powers of 2 go in cycles of length 3—the three numbers 2, 4, 1, over and over—while the powers of 3, 4, 5 and 6 repeat themselves in cycles of length 6, 3, 6 and 2 respectively. The key fact here is the these lengths—2, 3 and 6—all divide 6; it's because of this that the "x^6" column in Table 18-1 is all 1s. But is there a reason that all the lengths divide 6? What would the corresponding statement be for powers mod other primes? You may already have a clue as to the answer from having done Exercise 18.3.1, but to get more information let's examine another such table, for arithmetic mod 11.

18.4 More Tables

We start, naturally, by making a table of powers in arithmetic mod 11. This is not going to be too bad; there are a few tricks that'll reduce the computational work substantially. Let's begin by listing the first few powers of 2 (mod 11):

$$2^1 \equiv 2 \pmod{11}$$
$$2^2 \equiv 4 \pmod{11}$$
$$2^3 \equiv 8 \pmod{11}$$
$$2^4 \equiv 5 \pmod{11}$$
$$2^5 \equiv 10 \pmod{11}$$

Now, notice that $10 \equiv -1 \pmod{11}$, so that $2^6 \equiv -2$, $2^7 \equiv -2^2$ and so on: in other words, the next five powers of 2 will just be the negatives of the first five, right up to

$$2^{10} \equiv (2^5)^2 \equiv 1 \pmod{11}$$

where we can stop, having arrived at 1. We'll enter this in our table:

	x	x^2	x^3	x^4	x^5	x^6	x^7	x^8	x^9	x^{10}
2	2	4	8	5	10	9	7	3	6	1

Now for some tricks. Start with this: we know that $9 \equiv -2 \pmod{11}$; so if the powers of 2 (mod 11) are

$$2 \quad 4 \quad 8 \quad 5 \quad 10 \quad 9 \quad 7 \quad 3 \quad 6 \quad 1$$

the powers of 9 will be the same, but with a minus sign on the odd powers:

$$-2 \quad 4 \quad -8 \quad 5 \quad -10 \quad 9 \quad -7 \quad 3 \quad -6 \quad 1$$

or in other words

$$9 \quad 4 \quad 3 \quad 5 \quad 1 \quad 9 \quad 4 \quad 3 \quad 5 \quad 1$$

We'll enter this into our table:

	x	x^2	x^3	x^4	x^5	x^6	x^7	x^8	x^9	x^{10}
2	2	4	8	5	10	9	7	3	6	1
9	9	4	3	5	1	9	4	3	5	1

Now another trick: we've seen that

$$2 \times 2^9 \equiv 2^{10} \equiv 1 \pmod{11},$$

and it follows that $2^9 \equiv 6$ is the reciprocal of 2 (mod 11): multiplying by 6 is the same as dividing by 2. The "6" row of our power table will thus be just the "2" row backward: if the powers of 2 are

$$2 \quad 4 \quad 8 \quad 5 \quad 10 \quad 9 \quad 7 \quad 3 \quad 6 \quad 1$$

the powers of 6 will be

$$6 \quad 3 \quad 7 \quad 9 \quad 10 \quad 5 \quad 8 \quad 4 \quad 2 \quad 1.$$

We can then apply the -1 trick to this row: since $5 \equiv -6 \pmod{11}$, the "5" row will be the same as this one with minus signs on the odd powers. We'll enter both of these in our table:

	x	x^2	x^3	x^4	x^5	x^6	x^7	x^8	x^9	x^{10}
2	2	4	8	5	10	9	7	3	6	1
5	5	3	4	9	1	6	3	7	9	1
6	6	3	7	9	10	5	8	4	2	1
9	9	4	3	5	1	9	4	3	5	1

Now, to finish this off, let's try one more dodge. We know that $2^2 = 4$ (that much hasn't changed!), and it follows that *the powers of 4 are exactly the even powers of 2*: that is,

$$4^k = (2^2)^k = 2^{2k}.$$

In other words, we can write out the "4" row in our table of powers simply by taking every other entry in the "2" row: if the "2" row is

$$2 \quad 4 \quad 8 \quad 5 \quad 10 \quad 9 \quad 7 \quad 3 \quad 6 \quad 1$$

then the "4" row will be

$$4 \quad 5 \quad 9 \quad 3 \quad 1 \quad 4 \quad 5 \quad 9 \quad 3 \quad 1$$

Likewise, since $8 = 2^3$, the "8" row can be gotten by reading off every third entry in the "2" row:

$$8 \quad 9 \quad 6 \quad 4 \quad 10 \quad 3 \quad 2 \quad 5 \quad 7 \quad 1$$

Now, using these and the corresponding reciprocals and negatives, we can polish off our table: we arrive at the complete Table 18-2.

TABLE 18-2 Powers Mod 11

	x	x^2	x^3	x^4	x^5	x^6	x^7	x^8	x^9	x^{10}
1	1	1	1	1	1	1	1	1	1	1
2	2	4	8	5	10	9	7	3	6	1
3	3	9	5	4	1	3	9	5	4	1
4	4	5	9	3	1	4	5	9	3	1
5	5	3	4	9	1	6	3	7	9	1
6	6	3	7	9	10	5	8	4	2	1
7	7	5	2	3	10	4	6	9	8	1
8	8	9	6	4	10	3	2	5	7	1
9	9	4	3	5	1	9	4	3	5	1
10	10	1	10	1	10	1	10	1	10	1

Notice that the only powers we actually had to evaluate were the first five powers of 2! Bearing that in mind, do the following exercise:

Exercise 18.4.1 Work out the complete table of powers mod 13. You may use the same methods to avoid a lot of calculation (all you really need to compute are the first six powers of 2).

18.5 Fermat's Theorem

Take a look at the table of powers mod 11. As we observed in general, each row will simply repeat itself over and over, in cycles of length 10 (the "2", "6", "7" and "8" rows), or length 5 (the "3", "4", "5" and "9" rows), or length 2 (the "10" row) or 1 (the "1" row). Again, we see that all these lengths divide 10—in other words, the column of 10th powers consists entirely of 1s. If you did Exercise 18.4.1 for the powers mod 13, you probably observed a similar phenomenon: the rows repeat in cycles of lengths 12, 6, 4, 3, 2 or 1—all divisors of 12, so that the column of 12th powers consists entirely of 1s.

In other words, for each of the primes $p = 5$, 7, 11 and 13 we've investigated, it is the case that *the* $(p-1)^{\text{st}}$ *power of any nonzero number is equal to 1*. Can we conclude from our tables that this is always the case? No, certainly not—there are patterns in number theory that hold true for the first hundred or even thousand primes, but that fail in general. However, our data suggest that *this is a reasonable question to ask*. This is an important stage in mathematical knowledge, which might be viewed as the experimental stage, where we do many computations of a similar nature to try to determine *what is true*.

At some point, though, a mathematician will stop making tables and gathering data and will ask: how can we tell if this pattern holds for every prime p? How can we make one calculation, in other words, that will guarantee that if we take *any* prime number p, and *any* number a between 1 and $p-1$, that a^{p-1} will be congruent to 1 (mod p)?

This question was first asked (and answered) by Pierre Fermat. Fermat was a magistrate in Toulouse in the 17th century, at a time when there were few professional mathematicians. He was, however, a fairly active amateur; although he traveled no farther from home than to Bordeaux, he carried on a large correspondence with the scholars of his time, including Pascal and Huygens.

Fermat left almost no proofs of his results behind. (Indeed, a large part of Euler's work in number theory consisted in obtaining proofs of the statements in Fermat's collected work on the subject.) In particular, we don't know how Fermat analyzed this question, but here's a very clever argument that was found by Euler, some 100 years later. We'll carry it out first in a concrete example, and then in general, evaluating 2^6 (mod 7) and then a^{p-1} (mod p) for any prime number p and any nonzero a in arithmetic mod p.

So: say to begin with we want to evaluate 2^6 (mod 7); that is, take 6 copies of 2 and multiply them out, and see what we get (mod 7). (Of course, we could do this in five seconds on our calculator, but the idea is to find a way of calculating that we can carry out for any pair of numbers!) The trick is, we write the 2's in different forms: specifically, we write

$$2 \equiv \frac{2}{1} \equiv \frac{4}{2} \equiv \frac{6}{3} \equiv \frac{8}{4} \equiv \frac{10}{5} \equiv \frac{12}{6} \pmod 7$$

or, if we express the numerators (mod 7),

$$2 \equiv \frac{2}{1} \equiv \frac{4}{2} \equiv \frac{6}{3} \equiv \frac{1}{4} \equiv \frac{3}{5} \equiv \frac{5}{6} \pmod 7.$$

Note that these expressions all make sense, since the denominators are all nonzero. If we multiply these fractions together, we see that

$$2^6 \equiv \frac{2 \cdot 4 \cdot 6 \cdot 1 \cdot 3 \cdot 5}{1 \cdot 2 \cdot 3 \cdot 4 \cdot 5 \cdot 6} \pmod 7.$$

Now, the denominator of this fraction is obviously just the product of all the numbers between 1 and 6. But so is the numerator! And this is no accident: the numerator is just the product of the multiples of 2 from $1 \cdot 2$ to $6 \cdot 2$ in arithmetic mod 7—in other words, the entries of the "2" row in the multiplication table for arithmetic mod 7. But we've already seen that, in arithmetic mod 7, each row of the multiplication table (other than the "0" row) contains every number from 1 to 6 exactly once. Either way, we see that the numerator of this fraction equals the denominator, and since the two are nonzero the quotient is 1. Thus we arrive—admittedly, in considerably more time that it would have taken us just to multiply the thing out—at the conclusion that

$$2^6 \equiv 1 \pmod 7.$$

Now, as we said, the point of the last calculation is that we can carry it out in general. Let's do it: suppose we pick any prime number p and any number a between 1 and $p-1$. Again, we want to evaluate $a^{p-1} \pmod p$; that is, take $p-1$ copies of a and multiply them out, and see what we get $\pmod p$. As before, we write the a's in different forms: specifically, we write

$$a \equiv \frac{a}{1} \equiv \frac{2a}{2} \equiv \frac{3a}{3} \equiv \cdots \equiv \frac{(p-1)a}{p-1} \pmod p.$$

If we multiply these together, we see that

$$a^{p-1} \equiv \frac{a \cdot 2a \cdot 3a \cdot \cdots \cdot (p-1)a}{1 \cdot 2 \cdot 3 \cdot \cdots \cdot (p-1)} \pmod p.$$

Once more, the denominator of this fraction is just the product of all the numbers between 1 and $p-1$. The numerator, on the other hand, is just the product of the multiples of a from $1 \cdot a$ to $(p-1) \cdot a$—in other words, the entries of the "a" row in the multiplication table for arithmetic mod p. But we have seen that, in arithmetic mod any prime number p, *each row of the multiplication table contains every number from 1 to* $p-1$ *exactly once*: that is, the numbers

$$a, 2a, 3a, \ldots, (p-2)a, (p-1)a$$

appearing in the numerator are just the numbers

$$1, 2, 3, \ldots, p-2, p-1$$

appearing in the denominator, in another order. (Note also that the product of these numbers is not equal to 0 in arithmetic mod p: since none of the numbers $1, 2, 3, \ldots, p-1$ is divisible by p, their product can't be.) Since the order of multiplication doesn't matter, the numerator of this fraction is also just the product of all the numbers between 1 and $p-1$; that is, it equals the denominator and since the two are nonzero the quotient is 1.

Well, that does it: for any prime number p and any number $a \not\equiv 0 \pmod p$, we can carry out this calculation and see that $a^{p-1} \equiv 1 \pmod p$. We have thus formulated and proved a classical theorem in number theory, called *Fermat's Theorem*:

> If p is any prime number and a any number not divisible by p, then
>
> $$a^{p-1} \equiv 1 \pmod p.$$

There is another, equivalent, way to state Fermat's Theorem on powers, that doesn't even require the hypothesis that a is not divisible by p. Namely, if p is any prime number and a is any number:

$$a^p \equiv a \pmod p.$$

Indeed, if a is *not* divisible by p, this follows from the theorem by multiplying both sides of the identity $a^{p-1} \equiv 1 \pmod{p}$ by a. If a is divisible by p, so is a^p, and both sides of the formula are congruent to $0 \pmod{p}$.

Euler also gave an inductive argument to prove this version of the formula, using properties of the binomial coefficients! (If you need to review, the relevant facts are the binomial theorem, discussed in Section 6.4, and the divisibility properties of binomial coefficients worked out in Section 11.3.)

Euler starts by proving it for $a = 2$. We do this by applying the binomial theorem to the sum $1 + 1 = 2$: as we've seen, this says that

$$2^p = (1+1)^p = 1 + \binom{p}{1} + \binom{p}{2} + \cdots + \binom{p}{p-1} + 1.$$

Now, recall from Section 11.3 that for $k = 1, 2, \ldots, p-1$ *all of the binomial coefficients* $\binom{p}{k}$ *are divisible by* p. If we're working in arithmetic mod p, then, all the terms in the binomial theorem are zero except for the first and last, which gives us

$$2^p \equiv 2 \pmod{p}.$$

Next, let's prove the formula for $a = 3$. Once more we apply the binomial theorem, this time to the sum $1 + 2 = 3$: we see that

$$3^p = (1+2)^p = 1 + \binom{p}{1} \cdot 2 + \binom{p}{2} \cdot 2^2 + \cdots + \binom{p}{p-1} \cdot 2^{p-1} + 2^p.$$

As before, all the middle terms—from $\binom{p}{1} \cdot 2$ to $\binom{p}{p-1} \cdot 2^{p-1}$—are equal to zero (mod p) because the binomial coefficients are. What's more, we just saw that $2^p \equiv 2 \pmod{p}$. We conclude, then, that

$$3^p \equiv 1 + 2^p \equiv 3 \pmod{p}.$$

You can probably see how this is going to go: we can do 4^p next, then 5^p and so on. In general, suppose that we've managed to prove the formula $a^p \equiv a \pmod{p}$. We will show that $(a+1)^p \equiv a + 1 \pmod{p}$. To do this, we evaluate $(a+1)^p$ by the binomial theorem:

$$(a+1)^p = 1 + \binom{p}{1} a + \binom{p}{2} a^2 + \cdots + \binom{p}{p-1} a^{p-1} + a^p.$$

Again, all the terms involving a binomial coefficient $\binom{p}{k}$ for $k = 1, 2, \ldots, p-1$ are divisible by p. In arithmetic mod p this means

$$(a+1)^p \equiv 1 + a^p \pmod{p}.$$

But we've already seen that $a^p \equiv a \pmod{p}$, so we obtain the next formula

$$(a+1)^p \equiv a + 1 \pmod{p}$$

as desired. In other words, knowing the truth of the statement for $a = 2$ we may deduce the truth of the statement for $a = 3$; the truth of the statement for $a = 3$ implies the statement for $a = 4$, and so on. In some sense, this is an easier proof of Fermat's Theorem, but it uses very little of the multiplicative and power structure in arithmetic mod p.

We'll see in the next section a couple ways to use Fermat's theorem as a computational tool; and its generalization to arithmetic mod n for composite numbers n,

Euler's Theorem, will be the cornerstone of the coding theory we'll discuss later. But for the moment, we just want to pause and reflect on the process by which we arrived at this knowledge. In fact, this is typical of the way mathematicians operate: work out a lot of examples to come up with a (hypothetical) pattern; test this pattern in some more cases; and, once we're convinced that the pattern is truly universal, try to figure out a reason why it must be so, in the form of a general proof. If we can find two different proofs, as Euler did for Fermat's theorem, so much the better!

18.6 Calculating Powers mod p

Fermat's theorem, in the original or modified form, is not just an amusing pattern in the powers of a number in arithmetic mod p; it's also an extremely useful tool in modular arithmetic. To finish off this chapter, we'll look at some of the ways it can be used.

Probably the most common application is in computing powers. To see how this goes, let's revisit Exercise 18.2.1 from the second section of this chapter. The problem asks us to evaluate the power 11^{55} (mod 19), and the way we did it back then was straightforward: we simply found the powers 11^2, 11^4, 11^8, 11^{16} and 11^{32} (mod 19) by successive squaring, and then multiplied the relevant ones out to arrive at

$$11^{55} \equiv 11^{32} \times 11^{16} \times 11^4 \times 11^2 \times 11 \pmod{19}.$$

But there's a much better way, now that we know Fermat's Theorem! This tells us that for any nonzero number x in arithmetic mod 19

$$x^{18} \equiv 1 \pmod{19}.$$

Moreover, the same thing is true for x^{36}, x^{54} and so on: we see that

$$x^{36} \equiv (x^{18})^2 \equiv 1 \pmod{19},$$
$$x^{54} \equiv (x^{18})^3 \equiv 1 \pmod{19},$$

and similarly for x^{18k} for any whole number k. To put it another way, Fermat tells us that the powers of any number x in arithmetic mod 19 repeat over and over in cycles of length 18, or in other words

$$x^{k+18} \equiv x^k \pmod{19}.$$

In particular, in arithmetic mod p, *any power x^a of any nonzero number x whose exponent a is divisible by $p-1$ is congruent to 1 (mod p).*

In our present circumstances, we know that $11^{54} \equiv 1$ (mod 19), so we can write, simply,

$$11^{55} \equiv 11^{54} \times 11 \pmod{19}$$
$$\equiv 11 \pmod{19}.$$

Now, wasn't that easy? In fact, we can do this anytime we want to evaluate a power modulo a prime: according to Fermat's Theorem, we can throw away any multiple of $p-1$ from the exponent, in effect replacing the exponent we start with by its remainder under division by $p-1$. Here's another example:

EXAMPLE 18.6.1 Find 6^{373} (mod 11).

SOLUTION We don't want to evaluate this directly, even using the method of Section 18.2. Instead, we use Fermat's Theorem, which tells us that $6^{10} \equiv 1$ (mod 11), and hence that

$$\begin{aligned} 6^{370} &\equiv (6^{10})^{37} \pmod{11} \\ &\equiv 1 \pmod{11} \end{aligned}$$

and we can use this to write

$$6^{373} \equiv 6^{370} \times 6^3 \equiv 6^3 \pmod{11}$$

So all we have to do is evaluate 6^3 (mod 11), which is easy: $6^2 \equiv 36 \equiv 3$ (mod 11), so $6^3 \equiv 6 \times 3 \equiv 18 \equiv 7$ (mod 11), and this is our answer. ■

Of course, using Fermat's Theorem in calculating powers mod a prime p just reduces the exponent to a number less than p; it doesn't eliminate it entirely. In cases where p is large, in particular, we want to use this in conjunction with the method described in Section 18.2. Here's an example:

EXAMPLE 18.6.2 Find 7^{286} (mod 13).

SOLUTION We start by using Fermat's Theorem: we write

$$286 = 23 \times 12 + 10$$

and since we know that $7^{12} \equiv 1$ (mod 13), this means we can write

$$\begin{aligned} 7^{286} &\equiv (7^{12})^{23} \times 7^{10} \pmod{13} \\ &\equiv 7^{10} \pmod{13} \end{aligned}$$

Now we're left with the problem of finding 7^{10} (mod 13), which we do by the method of Section 18.2: we write

$$\begin{aligned} 7^2 &\equiv 49 \equiv 10 \pmod{13} \\ 7^4 &\equiv 10^2 \equiv 9 \pmod{13} \\ 7^8 &\equiv 9^2 \equiv 3 \pmod{13} \end{aligned}$$

so that we arrive at the answer:

$$7^{286} \equiv 7^{10} \equiv 7^8 \times 7^2 \equiv 10 \times 3 \equiv 4 \pmod{13}.$$ ■

Exercise 18.6.3 Evaluate the power 2^{83} (mod 23).

There's another way in which we can use Fermat's Theorem to help us calculate reciprocals.

Suppose, for example, that we want to find the reciprocal of a number—say, 8—in arithmetic mod 19. Of course, we already know ways of doing this, but here's another one: Fermat tells us that

$$8^{18} \equiv 1 \pmod{19}.$$

This means in particular that

$$8^{17} \times 8 \equiv 1 \pmod{19}$$

—in other words, 8^{17} is the reciprocal of 8 (mod 19)! We can evaluate this by the method of Section 18.2: we write

$$8^2 \equiv 64 \equiv 7 \pmod{19}$$
$$8^4 \equiv 7^2 \equiv 49 \equiv 11 \pmod{19}$$
$$8^8 \equiv 11^2 \equiv 7 \pmod{19}$$
$$8^{16} \equiv 7^2 \equiv 11 \pmod{19}$$

so $8^{17} \equiv 8 \times 8^{16} \equiv 8 \times 11 \equiv 12 \pmod{19}$ is the answer.

Exercise 18.6.4 Find the reciprocal $1/12$ in arithmetic mod 67 two ways: by the method of the preceding chapter, and by using Fermat's Theorem.

18.7 How Did Fermat Find His Theorem?

The simple answer is that we don't know. As we said, Fermat left very little, either of his motivation or of his arguments, behind him. In his day, mathematicians who found a new result or method often didn't publish it openly, but rather challenged others to solve problems requiring this knowledge. Fermat did make an attempt to persuade Pascal to help him write up some of his discoveries, but was unable to do so.

The lack of documentation has not stopped mathematical historians from speculating on the problems motivating Fermat. One was undoubtedly the search for an easily described sequence of prime numbers. One of Fermat's contemporaries was a traveling priest, Mersenne, who proposed that the sequence $2^2 - 1, 2^3 - 1, 2^4 - 1, \ldots$ should contain infinitely many prime numbers. Since

$$2^{ab} - 1 = (2^a - 1)(2^{a(b-1)} + 2^{a(b-2)} + \cdots + 2^a + 1)$$

the number $2^n - 1$ can only be prime when n itself is a prime p. The first few terms in this sequence

$$2^2 - 1 = 3, \quad 2^3 - 1 = 7, \quad 2^5 - 1 = 31 \quad \text{and} \quad 2^7 - 1 = 127$$

are indeed prime, but the next term, $2^{11} - 1 = 2047$, factors as 23×89. It became a question, then, given a prime number p, to determine if the Mersenne number $2^p - 1$ is indeed a prime. By Fermat's time, mathematicians were grappling with the number

$$2^{37} - 1 = 137{,}438{,}953{,}471,$$

unable to say whether or not it was prime. Fermat resolved the issue, showing that this number was *not* prime, by giving its prime factorization: $137{,}438{,}953{,}471 = 223 \times 616{,}318{,}177$.

To find these factors Fermat probably used his theorem on powers, and indeed it may have been a motivation in the theorem's discovery. Here's how Fermat may have approached the problem: suppose a prime number p divides $2^{37} - 1$. Then we have

$$2^{37} \equiv 1 \pmod{p}.$$

On the other hand, by Fermat's theorem:

$$2^{p-1} \equiv 1 \pmod{p}.$$

Now, consider these two statements. How can they both be true? Well, one way is if 37 divides $p - 1$—then 2^{p-1} would be a power of 2^{37}, and the first statement would imply the second. In fact, that's the only way both statements can be true. To see this, suppose that 37 didn't divide $p - 1$. In that case, 37 and $p - 1$ would be relatively prime (37 is a prime number), and by Euclid's algorithm, there would be a combination $37a + (p - 1)b = 1$. Then

$$2 \equiv 2^1 \equiv 2^{37a+(p-1)b} \equiv 1 \pmod{p},$$

which is false. Hence 37 must divide $p - 1$.

We see from this that if a prime number p divides $2^{37} - 1$, then p must be of the form $p = 37k + 1$ for some whole number k. Since p is odd, moreover, k must be even. Thus Fermat was able to narrow down his search for a prime factor of $2^{37} - 1$ to a relatively small list of candidates. In fact, the first few values $k = 2, 4, 6$ give the potential divisors

$$2 \times 37 + 1 = 75,$$
$$4 \times 37 + 1 = 149$$

and

$$6 \times 37 + 1 = 223;$$

and the last number is the small prime factor of $2^{37} - 1$ that Fermat himself found!

Nowadays, it's known that the primes p less than 1000 for which $2^p - 1$ is a prime are

$$p = 2, 3, 5, 7, 13, 19, 31, 61, 89, 107, 127, 521 \text{ and } 607.$$

We expect that there are infinitely many Mersenne primes, but there are only 39 known to date.

Chapter 1

Roots

Multiplication is built into the very definition of a number system: it's part of what it means to have a number system, as we defined it in Section 14.3, that you can multiply any two numbers. As we've seen, though, the inverse operation, division, is more problematic; we may or may not be able to divide one number by another in a given number system, and even when we know that it's possible carrying out the calculation may be tricky.

Taking powers of a number, and the reverse operation, taking roots, have a similar relationship to each other. Since we can always multiply numbers, we can always square, cube or take any power of a given number. But we can't always find a square root or cube root or k^{th} root of a given number; and, correspondingly, even in cases where we know it's possible it may be difficult to actually find the root.

In this chapter we'll investigate roots in arithmetic mod p: when they exist, and how we can find them when they do. We'll proceed very much as in the chapter on division: we'll first ask the question of when a number has a root, and look at examples to try and guess a general answer to that question. Having done that, we'll go on and show you how to calculate roots in practice.

One note before we start: as in the case of powers, in this initial chapter we'll focus exclusively on the case of arithmetic mod p, where p is a prime number. There are two reasons for this. First, as you might expect, the situation is just simpler in that case, and it'll be easier if we get our bearings there before we move on to the more complex case of composite moduli. The other reason is more absolute: as we'll see, the key to taking roots mod p—both to saying when they exist, and to finding them—is Fermat's Theorem, which only works in arithmetic mod a prime number p. Our program, accordingly, is this: we'll start by showing you how to take roots in arithmetic mod a prime number p. Then in Chapter 20 we'll go back to Fermat's Theorem, and investigate its analog for composite moduli. Finally, once we've done that, we'll be able to complete our discussion of powers and roots in any modular number system.

.1 What Do We Mean by a kth Root?

By the square of a number, we mean that number multiplied by itself; by the square root of a number a, we mean a number whose square is a—if one exists. Likewise, by the cube root of a number a, we mean a number x such that $x^3 = a$, again if one exists. By the kth root of a number a, we mean a number whose k^{th} power is equal to a; that is, a solution of the equation

$$x^k = a,$$

as always, if one exists.

Before we launch into our discussion of roots in modular arithmetic, let's survey the situation in ordinary arithmetic. In the realm of whole numbers or fractions, as we've seen, roots rarely exist: in order for a number to have a square root, for example, all the exponents in its prime factorization have to be even; to have a cube root, all the exponents have to be divisible by 3, and so on.

If we enlarge our number system to include all decimal numbers, the situation is a good bit more regular. Any positive decimal number will have a square root, and every decimal number has a cube root; in general, if k is an odd number, every decimal number has a kth root, while if k is even then a decimal number will have a kth root whenever it's positive. It's also worth remarking that in the system of complex numbers, every nonzero number has exactly k different k^{th} roots.

One other thing to notice is that there may be more than one root of a number. You're used to this in ordinary arithmetic, where by "the square root of 4" we could mean either 2 or -2. In fact, in ordinary number systems, if a square root exists then there'll be two: x and $-x$, of which one will be positive and one negative; we typically resolve this ambiguity by choosing the positive root. You should realize, though, that in modular arithmetic there are no positive and negative numbers; correspondingly, when we do see similar ambiguity in roots mod p there will be no natural choice among them.

19.2 Do Roots Exist, and If So How Many?

When we started our discussion of division, asking when quotients existed in modular arithmetic, our first step was to dust off the multiplication tables for arithmetic mod n and examine them to answer that question in specific cases. By analogy, we'll do the same thing here: to answer the question of when we can find roots mod p, we'll start by examining the tables of powers we created in the last chapter. Let's start with the table of powers mod 7:

	x	x^2	x^3	x^4	x^5	x^6	x^7	x^8	x^9	x^{10}	x^{11}	x^{12}	x^{13}
1	1	1	1	1	1	1	1	1	1	1	1	1	1
2	2	4	1	2	4	1	2	4	1	2	4	1	2
3	3	2	6	4	5	1	3	2	6	4	5	1	3
4	4	2	1	4	2	1	4	2	1	4	2	1	4
5	5	4	6	2	3	1	5	4	6	2	3	1	5
6	6	1	6	1	6	1	6	1	6	1	6	1	6

From this table we should be able to tell whether or not a number has a root (mod 7). For example, we could ask: does 5 have a square root (mod 7)—in other words, is there a number whose square is 5 (mod 7)? To answer this, we look at the "x^2" column of our table, where all the squares (mod 7) appear; 5 is not in that column, from which we conclude that 5 has no square root (mod 7). Similarly, if we asked whether 2 has a square root (mod 7) we'd look for 2s in the "x^2" column of our table. In this case we'd see a couple, in the "3" row and the "4" row, meaning that 3^2 and $4^2 \equiv 2 \pmod 7$. We conclude that 3 and 4 are square roots of 2 in arithmetic mod 7.

Exercise 19.2.1 Using the power table above, say whether or not the following roots exist in arithmetic mod 7; and if they do, say how many and find them all.

1. $\sqrt[3]{5}$
2. $\sqrt[3]{6}$
3. $\sqrt[5]{2}$
4. $\sqrt[5]{3}$
5. $\sqrt[13]{2}$

Before we move on, let's look for patterns in this table. In particular, we should ask one question: for which k is it the case that *every* number in arithmetic mod 7 has a kth root? For example, looking at the table we see that not every number has a square root, or a cube root, or a fourth root in arithmetic mod 7. On the other hand, it *is* true that every number has a 5th root: the "x^5" column contains every nonzero number in arithmetic mod 7. Likewise we see that the "x^7" and "x^{11}" columns both include every nonzero number, meaning that every number has a 7th root and an 11th root. (Of course, this is really not news: Fermat's Theorem says that the "x^7" and "x^{11}" columns are just repeats of the "x" and "x^5" columns, that is, the 11th root of a number (mod 7) is the same as its 5th root.) In other words, the answer to our question, "Can we always take a kth root in arithmetic mod 7?" is "Yes" whenever k is 5, 7, 11, 13, 17, 19 and so on.

Let's move on to arithmetic mod 11. Again, we'll drag out the old power table:

	x	x^2	x^3	x^4	x^5	x^6	x^7	x^8	x^9	x^{10}
1	1	1	1	1	1	1	1	1	1	1
2	2	4	8	5	10	9	7	3	6	1
3	3	9	5	4	1	3	9	5	4	1
4	4	5	9	3	1	4	5	9	3	1
5	5	3	4	9	1	6	3	7	9	1
6	6	3	7	9	10	5	8	4	2	1
7	7	5	2	3	10	4	6	9	8	1
8	8	9	6	4	10	3	2	5	7	1
9	9	4	3	5	1	9	4	3	5	1
10	10	1	10	1	10	1	10	1	10	1

As in the previous case, we can use this table to find roots in arithmetic mod 11. But we're more interested in asking the qualitative question: *for which numbers k is it true that we can always take k^{th} roots in arithmetic mod* 11*?* For example, we see that not every number has a square root (mod 11), but it is true that every number has a cube root; it's not true that every number has a 4th, 5th or a 6th root, but it is true that every number has a 7th root. In fact, looking at the table and asking which columns contain all the numbers 1, 2, 3, ... ,10, the answer is: the "x" column; the "x^3" column; the "x^7" column and the "x^9" column. This means that *we can always take a* k^{th} *root in arithmetic* mod 11 *whenever k is* 1, 3, 7, 9, 11, 13, 17, 19 *and so on.*

Exercise 19.2.2 Examine the power tables you worked out in Exercises 18.3.1 and 18.4.1, and use them to answer the questions: for which k is it the case that every number has a kth root in arithmetic mod 5? In arithmetic mod 13?

Now that we have some data, let's see if we can spot the pattern, and answer in general the question of when we can take k^{th} roots in arithmetic mod p. First, let's tabulate what we know: we'll make a chart listing the primes we've investigated so far, and next to each the orders of the roots we know exist:

prime p	*numbers k such that we can always find* k^{th} *roots mod p*
5	1, 3, 5, 7, 9, 11, 13, 15, ...
7	1, 5, 11, 13, 17, 19, 23, ...
11	1, 3, 7, 9, 11, 13, 17, 19, 21, ...
13	1, 5, 7, 11, 13, 17, 19, 23, 25, ...

Now, stop. Before you look ahead to see the answer, you should pause and try out your pattern recognition skills: *what property characterizes the numbers to the right of each prime in the left-hand column?*

STOP.

CLOSE THE BOOK.

GRAB A PAD OF PAPER AND A PEN.

WORK OUT SOME EXAMPLES ON YOUR OWN.

THINK.

19.3 The Answer

The answer is, *the numbers to the right of each prime p are exactly the numbers relatively prime to* $p - 1$. What's more, this pattern holds true in general, as we'll see in the next section. We haven't proved it yet, but we'll state it here:

> If p is any prime, and k is any number relatively prime to $p - 1$, then every number has a unique k^{th} root in arithmetic mod p.

Exercise 19.3.1 Check out the boxed statement in the case $p = 17$ and $k = 2$ and 3: that is, verify that not every number has a square root (mod 17), but every number does have a cube root (mod 17).

Our goal now is twofold: we want to show that the boxed statement is always true; and, in cases where the boxed statement says that a kth root of a number exists (mod p), we want to develop techniques for finding it. Those two goals are very much interdependent: as is usually the case in mathematics, our understanding of mathematical phenomena grows out of our techniques, and our techniques are rooted in our understanding. In fact, the investigation of the next section is a great example of how we often develop a technique and prove a theorem hand in hand.

19.4 Roots Are Powers; Powers, Roots

The first thing we should recognize is that the boxed statement is fundamentally counterintuitive. Think about the situation with whole numbers or fractions: if we look at the whole numbers between 1 and 1,000,000, for example, and ask how many have square roots, cube roots and so on we find that

- 1,000 have square roots
- 100 have cube roots
- 31 have fourth roots
- 15 have fifth roots
- 10 have sixth roots
- 7 have seventh roots

—in other words, roots are getting sparser and sparser as the order of the root increases.

But what we seem to be seeing is that arithmetic mod p behaves quite differently: mod 13, for example, every number is a 7th power, and an 11th power, and so on. If we want to prove that this is the case, and also if we want to develop a technique for finding those roots, we have to understand why this is true.

The answer has to do with the cyclic pattern of powers mod p: instead of the powers of a given number wandering off to infinity, according to Fermat's Theorem they repeat themselves forever in cycles of length $p - 1$. One aspect of this in particular is that in arithmetic mod p *every number is a power of itself.* In fact, by Euler's version of the theorem

$$a^p \equiv a \pmod{p}.$$

This cyclic pattern is exactly what we have to use, both to prove the boxed statement above and to actually calculate roots.

Let's see how this plays out in an example. Suppose we want to find a fifth root $\sqrt[5]{2}$ of 2 (mod 17). The boxed statement tells us there is one, since 5 is relatively prime to 16, but doesn't give us a clue how to find it. Of course, if we wanted to grind it out we would just make a list of the fifth powers of numbers mod 17—write out the "x^5" column of the power table in arithmetic mod 17, in other words—until we found a "2," but Fermat's Theorem provides a better way. It tells us that $2^{16} \equiv 1 \pmod{17}$, but it also tells us more: it tells us that

$$2^{17} \equiv 2^{16} \times 2 \equiv 2 \pmod{17},$$

and likewise

$$2^{33} \equiv (2^{16})^2 \times 2 \equiv 2 \pmod{17},$$

and

$$2^{49} \equiv (2^{16})^3 \times 2 \equiv 2 \pmod{17},$$

and so on. In fact, it tells us that we can write 2 as a power of itself in many ways:

$$2 \equiv 2^{17} \equiv 2^{33} \equiv 2^{49} \equiv 2^{65} \equiv 2^{81} \equiv \dots \pmod{17}.$$

Now, of all these different expressions for 2 in arithmetic mod 17, there is one that is useful to us in our present circumstance of looking for a fifth root of 2. The one that does it for us in this case is

$$2^{65} \equiv 2 \pmod{17},$$

because *65 is a multiple of 5*: the fact that $65 = 5 \times 13$ means we can rewrite 2^{65} as $(2^{13})^5$, so that

$$(2^{13})^5 \equiv 2 \pmod{17}.$$

In other words, the fifth power of 2^{13} is 2; so *2^{13} is a fifth root of 2 in arithmetic mod 17!*

Excellent! What this seems to suggest is that, by virtue of the cyclical nature of the powers of a given number in arithmetic mod p, we can undo the process of taking powers of a number by taking further powers! (After all, if we move along a circle, we can go back by going forward.) Of course, if we want a numerical answer, we have to evaluate 2^{13} in arithmetic mod 17, but we already know how to do that.

Let's do one more numerical example before we describe the general method. Suppose we're working in arithmetic mod 29, and we're asked to find the cube root of 11. Once more, we use the only tool we have: we use Fermat's Theorem to reexpress 11 as a power of itself in different ways. Specifically, Fermat tells us that the powers of 11 in arithmetic mod 29 repeat in cycles of length 28, so that

$$11 \equiv 11^{29} \equiv 11^{57} \equiv 11^{85} \equiv \dots \pmod{29}.$$

Now, the key equality here is $11 \equiv 11^{57} \pmod{29}$, because the exponent 57 is divisible by 3: the fact that $57 = 19 \times 3$ allows us to write

$$11^{57} \equiv (11^{19})^3 \pmod{29}$$

so that

$$(11^{19})^3 \equiv 11 \pmod{29}$$

and hence 11^{19} is a cube root of 11 in arithmetic mod 29. Lastly, if we want an answer in the form of a number between 0 and 29, we evaluate 11^{19} in arithmetic mod 29: we write

$$11^2 \equiv 121 \equiv 5 \pmod{29}$$
$$11^4 \equiv 25 \equiv -4 \pmod{29}$$
$$11^8 \equiv 16 \pmod{29}$$
$$11^{16} \equiv 256 \equiv 24 \pmod{29}$$

Since $19 = 16 + 2 + 1$ this gives us

$$\begin{aligned} 11^{19} &\equiv 11^{16} \cdot 11^2 \cdot 11 \\ &\equiv 24 \cdot 5 \cdot 11 \\ &\equiv 15, \end{aligned}$$

so that's our answer: 15 is the cube root of 11 in arithmetic mod 29. Finally, if we're really doing this properly, we should double-check: if we work it out, we see that

$$15^3 \equiv 3{,}375 \equiv 11 \pmod{29}.$$

It's time to switch from numbers to letters, and see how this might go in general. Suppose we're looking for the kth root of a number a in arithmetic mod p. Fermat's Theorem tells us that we can write a as a power of itself: since $a^{p-1} \equiv 1 \pmod{p}$, we can keep multiplying a by a^{p-1} to get a again:

$$a \equiv a^p \equiv 2^{2p-1} \equiv 2^{3p-2} \equiv 2^{4p-3} \equiv \ldots \pmod{p}.$$

In other words, for any number ℓ,

$$a^{\ell(p-1)+1} \equiv (a^{p-1})^{\ell} \times a \equiv a \pmod{p}.$$

How does this help us find a kth root of a? Well, in the first case we worked out above, what saved us was that among the powers of 2 that were equal to 2 itself, there was one—2^{65}—whose exponent was divisible by 5. In the general setting we want to find a power of a that's congruent to a, and whose exponent is divisible by k. Now, the powers of a that Fermat's Theorem tells us are congruent to $a \pmod{p}$ are the powers of the form $a^{\ell(p-1)+1}$. So what this really means is that we want to find a number that is simultaneously of the form mk—that is, a multiple of k—and of the form $\ell(p-1)+1$—that is, congruent to $1 \pmod{p-1}$. In other words, we want to solve the equation

$$mk = \ell(p-1) + 1;$$

or, to put it another way, we want to find the reciprocal m of $k \pmod{p-1}$. Now, the good news: as we saw in Section 9.3, as long as k is relatively prime to $p-1$, we know how to do this!

Once we've done this, moreover, we're set: since $mk \equiv 1 \pmod{p-1}$, we know from Fermat that

$$a^{mk} \equiv a \pmod{p}.$$

Now we just rewrite a^{mk} as $(a^m)^k$; so this says that

$$(a^m)^k \equiv a \pmod{p};$$

or simply that a^m is the kth root of a (mod p). In other words, if k and m are reciprocals (mod $p-1$)—that is, if $mk \equiv 1$ (mod $p-1$)—then the kth root of a number is the same as its mth power.

Here's a summary of the process:

How to Find $\sqrt[k]{a}$ (mod p)

- Check that p is a prime, that a is nonzero (mod p), and that k is relatively prime to $p-1$. If not, this process won't work, and indeed there may not be a kth root of a (mod p).
- Use the Euclidean algorithm to find whole numbers m and ℓ such that

$$mk = \ell(p-1) + 1$$

as we saw how to do in Section 9.3; equivalently, find the reciprocal m of k in arithmetic mod $p-1$ as described in Section 17.3. Having done this, we know that

$$a^{mk} \equiv a^{\ell(p-1)+1} \equiv (a^{p-1})^{\ell} \cdot a \equiv a \pmod{p};$$

which means that

- $(a^m)^k \equiv a$ (mod p). In other words, the mth power of a is simultaneously the kth root of a (mod p). Thus all we have to do is
- Evaluate a^m (mod p); this is your answer. There's one last step:
- Check your answer by raising it to the kth power and seeing that you do indeed get a. This isn't officially part of the process, but it never hurts to make sure!

In fact, we've accomplished both our goals: not only have we shown that roots will always exist in arithmetic mod p whenever the order of the root is relatively prime to $p-1$, but we've found a way of calculating them.

Let's try another example, to see how it goes in practice:

EXAMPLE 19.4.1 Find the 7th root of 3 in arithmetic mod 19.

SOLUTION The first step is to check that 19 is indeed a prime (it is), that 3 is relatively prime to 19 (ditto) and that 7 is relatively prime to $19-1=18$ (ditto again).

The second step is to find the reciprocal of 7 (mod 18)—that is, we want to express 1 as a combination of 7 and 18. We run the Euclidean algorithm:

$$18 = 2 \times 7 + 4$$
$$7 = 1 \times 4 + 3$$
$$4 = 1 \times 3 + 1$$

and then we run it backwards to express 1 as a combination of 7 and 18:

$$\begin{aligned} 1 &= 4 - 3 \\ &= 4 - (7 - 4) \\ &= 2 \times 4 - 7 \\ &= 2 \times (18 - 2 \times 7) - 7 \\ &= 2 \times 18 - 5 \times 7. \end{aligned}$$

In other words, the reciprocal of 7 (mod 18) is -5, or 13. Check this:

$$7 \times 13 = 91 = 5 \times 18 + 1$$

According to the third step, then, the answer will be 3^{13} (mod 19). Let's evaluate this: we write

$$13 = 8 + 4 + 1$$

and make a list of powers of 3 (mod 19) by successive squaring:

$$\begin{aligned} 3^2 &\equiv 9 \pmod{19} \\ 3^4 &\equiv 9^2 \equiv 5 \pmod{19} \\ 3^8 &\equiv 5^2 \equiv 6 \pmod{19} \end{aligned}$$

to arrive at

$$\begin{aligned} 3^{13} &\equiv 3 \times 3^4 \times 3^8 \pmod{19} \\ &\equiv 3 \times 5 \times 6 \pmod{19} \\ &\equiv 14 \pmod{19}. \end{aligned}$$

Our conclusion, then, is that

$$\sqrt[7]{3} \equiv 14 \pmod{19}.$$

Now, let's not forget the fourth step. We know we should have the right answer, since

$$\begin{aligned} (3^{13})^7 &\equiv 3^{91} \pmod{19} \\ &\equiv (3^{18})^5 \times 3 \pmod{19} \\ &\equiv 3 \pmod{19} \end{aligned}$$

but it never hurts to check by actually taking the 7th power of 14 (mod 19) and seeing what we get. We write

$$7 = 4 + 2 + 1$$

and start squaring:

$$\begin{aligned} 14^2 &\equiv (-5)^2 \equiv 6 \pmod{19} \\ 14^4 &\equiv 6^2 \equiv 17 \equiv -2 \pmod{19}. \end{aligned}$$

We have, then

$$\begin{aligned} 14^7 &\equiv 14 \times 6 \times (-2) \pmod{19} \\ &\equiv 3 \pmod{19}. \end{aligned}$$

So 14 really is the 7th root of 3 (mod 19). ■

To sum up: using the cyclical nature of powers in modular arithmetic, we've found a systematic way of taking roots in arithmetic mod p by taking powers. What a world! And this is more than a theoretical advance: though the mathematicians who discovered these phenomena were interested only in elucidating the behavior of numbers, it turns out that the material of this chapter has fundamentally important applications.

In fact, this way of taking roots in modular arithmetic is at the heart of the modern codes that make it possible to transact business securely over networks: every time you give out your credit card number online, your computer encrypts that information by a process that relies on the ideas of this chapter. In the next chapter we'll extend all of this to the case of arithmetic mod a nonprime number n—we'll describe the analog of Fermat's Theorem in this case, and show how we can similarly use it to find roots in arithmetic mod n—and that will conclude this part of the book. In the last part, we'll take a break from all these numbers and discuss coding theory in general; then we'll be able to describe this encryption method in detail.

We do want to reiterate a warning, though: the process we've described here for finding the kth root $\sqrt[k]{a}$ of a number a in arithmetic mod p works only subject to three hypotheses:

(a) p must indeed be prime;

(b) a must be relatively prime to p; and

(c) k must be relatively prime to $p - 1$.

If the last of these conditions fails, the method simply won't work; when you try to find the reciprocal of k in arithmetic mod $p - 1$, the Euclidean Algorithm will tell you there isn't one. But a worse fate may befall you if the first condition isn't satisfied: if p isn't in fact prime, then Fermat's Theorem will probably be false. In this case the process may well proceed without a hitch—and yield the wrong answer.

Exercise 19.4.2 We don't mean to be malicious, but to emphasize the importance of checking the hypotheses before proceeding with the method described here, only two of the following four problems can be solved with the method of this chapter. Say which they are, and do them.

1. Find the fifth root $\sqrt[5]{3}$ of 3 in arithmetic mod 23
2. Find the fifth root $\sqrt[5]{7}$ of 7 in arithmetic mod 31
3. Find the cube root $\sqrt[3]{4}$ of 4 in arithmetic mod 57
4. Find the cube root $\sqrt[3]{4}$ of 4 in arithmetic mod 59

I'M NOT GOING TO DO MY MATH HOMEWORK.
© 1992 Watterson/Distributed by Universal Press Syndicate

LOOK AT THESE UNSOLVED PROBLEMS. HERE'S A NUMBER IN MORTAL COMBAT WITH ANOTHER. ONE OF THEM IS GOING TO GET SUBTRACTED, BUT WHY? HOW? WHAT WILL BE LEFT OF HIM?

IF I ANSWERED THESE, IT WOULD KILL THE SUSPENSE. IT WOULD RESOLVE THE CONFLICT AND TURN INTRIGUING POSSIBILITIES INTO BORING OL' FACTS.

I NEVER REALLY THOUGHT ABOUT THE LITERARY QUALITIES OF MATH.
I PREFER TO SAVOR THE MYSTERY.

Chapter 20

Euler's Theorem

We've come a long way in modular arithmetic: starting with the basic definitions, we've developed a fair degree of facility with calculation mod n. We have, in fact, just one more basic tool to introduce, and then we'll be able to discuss an application to coding theory. At that point, we'll be done with the topic—though in fact there are still a universe of questions to explore, many of which remain as the subject of current mathematical research.

20.1 Extending Fermat's Theorem

That one further tool is *Euler's Theorem*, and it doesn't even represent a new idea: it's simply the extension of Fermat's Theorem to the case of arithmetic mod n, where n need not be a prime number. Fermat's Theorem tells us that if p is a prime number, then the powers of any nonzero number a in arithmetic mod p repeat themselves in cycles of length $p - 1$. In particular, this says that $a^{p-1} \equiv a^0 \equiv 1 \pmod{p}$. Now we'd like a similar statement in arithmetic mod n: one that will tell us that a certain power of a number a is congruent to $1 \pmod{n}$.

There is one remark we should make before we start in. Going back to ordinary whole-number arithmetic, recall that every whole number may be uniquely factored into a product of primes. To say that two numbers a and n are relatively prime means that no prime number divides both of them. But the primes that divide any power of a are exactly the same primes as divide a. What this means is that if a number a is relatively prime to a number n, then so are all its powers; and, conversely, *if any power of a is relatively prime to n then a itself must be.*

In particular, if some power of a number a is congruent to $1 \pmod{n}$, then, since 1 is clearly prime to n, a itself must be relatively prime to n. Thus, any extension of Fermat's Theorem to arithmetic mod n—that is, any statement that a certain power of a number a will be $\equiv 1 \pmod{n}$—can only apply to numbers a that are relatively prime to n. This is, in a sense, not a new restriction: it's just the natural generalization to nonprime modulus n of the requirement in Fermat's Theorem that a not be divisible by p.

With this said, how should we go about investigating a possible extension of Fermat's Theorem? There are two ways, and in fact the contrast in methods is interesting in its own right. First, there is the sort of method we've been using throughout this part of the book: gather data from arithmetic mod n for a few relatively small choices of n, and try to see a pattern in this data. There is also another approach, less familiar to us in this text but very common in mathematics. We can go back to the proof of Fermat's Theorem, and try to analyze that: where does it use the fact that the modulus p is prime, and how could we modify it to remove its dependence on this assumption?

In the next two sections we'll try both approaches. This may seem a little redundant; if you want, you could skip one or the other. But it may also be interesting to see how we can be led to the same conclusion by two completely different lines of thought.

20.2 More #$%& Tables

We'll try the familiar approach first: gather data, and try to extrapolate. What this means in the present case is: we choose a modulus n, not a prime. We then list all the numbers a from 1 to $n - 1$ that are relatively prime to n, and make a table of their powers in arithmetic mod n as we did in Chapter 18. What we hope to see is a column of that table that consists entirely of 1s; we then want to see if we can guess a rule for which column it is.

Let's start with the case of arithmetic mod 8. This is particularly easy: there are four numbers between 1 and 7 that are relatively prime to 8—1, 3, 5 and 7—and as we see the minute we start taking powers, all four have squares congruent to 1 (mod 8). This makes for a singularly uninteresting table, but we'll give it anyway:

Table 20-1 Powers Mod 8

	x	x^2	x^3	x^4	x^5
1	1	1	1	1	1
3	3	1	3	1	3
5	5	1	5	1	5
7	7	1	7	1	7

In fact, we could have stopped after the second column: what we see is that all the even columns will consist of 1s, and correspondingly the odd columns will just be repeats of the first column.

Next, let's do the case of arithmetic mod 10. To begin with, there are four numbers in arithmetic mod 10 that are relatively prime to 10: 1, 3, 7 and 9. Making a table of their powers is easy: powers of 1 are all 1, while powers of 9 just alternate 1s and -1s; similarly, powers of 3 and 7 differ only in a minus sign in odd columns. Here we go:

Table 20-2 Powers Mod 10

	x	x^2	x^3	x^4	x^5
1	1	1	1	1	1
3	3	9	7	1	3
7	7	9	3	1	7
9	9	1	9	1	9

Again, we do see a column of all 1s: it's the fourth column, meaning that $a^4 \equiv 1 \pmod{10}$ for any a that is relatively prime to 10; and of course the same is true for a^8, a^{12} and any power a^{4k} divisible by 4.

The next case we will work out is $n = 14$. The numbers relatively prime to 14 are 1, 3, 5, 9, 11 and 13, and again it's not that much work to make a table of powers. We start with 3 and list its powers mod 14:

$$3^1 \equiv 3 \pmod{14}$$
$$3^2 \equiv 9 \pmod{14}$$
$$3^3 \equiv 27 \equiv 13 \pmod{14}$$
$$3^4 \equiv 39 \equiv 11 \pmod{14}$$
$$3^5 \equiv 33 \equiv 5 \pmod{14}$$
$$3^6 \equiv 15 \equiv 1 \pmod{14}$$

That gives us one row of our table; and we can read off the other rows from this: for example, since $9 = 3^2$, the powers of 9 are just the even powers of 3; since 5 is the reciprocal of 3, its powers are the powers of 3 in reverse order, and so on. The result is shown in Table 20-3. Once more, there is indeed a column of 1s; it's the sixth column.

TABLE 20-3 Powers Mod 14

	x	x^2	x^3	x^4	x^5	x^6	x^7
1	1	1	1	1	1	1	1
3	3	9	13	11	5	1	3
5	5	11	13	9	3	1	5
9	9	11	1	9	11	1	9
11	11	9	1	11	9	1	11
13	13	1	13	1	13	1	13

One more: let's do $n = 18$. The numbers relatively prime to 18 are 1, 5, 7, 11, 13 and 17, and we'll give you the results in Table 20-4.

TABLE 20-4 Powers Mod 18

	x	x^2	x^3	x^4	x^5	x^6	x^7
1	1	1	1	1	1	1	1
5	5	7	17	13	11	1	5
7	7	13	1	7	13	1	7
11	11	13	17	7	5	1	11
13	13	7	1	13	7	1	13
17	17	1	17	1	17	1	17

Now, it's your turn to make some tables.

Exercise 20.2.1 For each number n, make a list of the numbers a between 1 and $n-1$ that are relatively prime to n, and a table of their powers mod n. Is there a column consisting entirely of 1s? Which column is it?

1. $n = 9$

2. $n = 12$

3. $n = 15$

Now it's time to examine the evidence and try to answer the two questions we started out with. Is it always the case that we have a column of 1s? Well, the evidence is that the answer is yes: certainly it's true in every case we tried. Which column is it? Well, that might be a little trickier. In fact, we'll leave you to contemplate the information we have—we'd suggest you make a list of the numbers n we've tried, and next to each which columns in the table of powers mod n consists entirely of 1s—and scratch your head over that for a while. In any event, we'll have the answer by the end of the next section.

20.3 Pure Thought

There is a style of mathematics that absolutely abjures the sort of experimental approach we tried in the last section: it holds that we should be able to arrive at the truth by "pure thought." In this case, that might mean going back to the proof of Fermat's Theorem in Section 18.5 and seeing what it might tell us in the case of arithmetic modulo a composite number n.

Actually, we had two proofs of Fermat's Theorem, and it's instructive to see how well each adapts to the new world of nonprime modulus. We'll recall each briefly now, and see if either can be modified to fit the more general circumstances; we'll see that one fails, but one succeeds.

20.3.1 The Proof that Fails

Let's start with the proof using properties of the binomial coefficients. If you remember, this argument starts by showing that $2^p \equiv 2 \pmod{p}$, using the binomial expansion

$$2^p = (1+1)^p = 1 + \binom{p}{1} + \binom{p}{2} + \cdots + \binom{p}{p-1} + 1.$$

The key step in this argument is the observation that *all of the binomial coefficients* $\binom{p}{k}$ *for* $k = 1, 2, \ldots, p-1$ *are divisible by* p; this is what allowed Euler to eliminate all but the first and last terms mod p and deduce the formula. Now, to show that $\binom{p}{k}$ is divisible by p, we look at the formula:

$$\binom{p}{k} = \frac{p(p-1)(p-2)\ldots(p-k+1)}{k(k-1)(k-2)\ldots 1}.$$

We note that p appears once in the numerator, and does not divide any of the factors in the denominator. Thus the numerator is divisible by p, while the denominator is not. Now, if we have a quotient

$$q = \frac{a}{b}$$

then we also have

$$a = b \cdot q.$$

If a prime divides a, then, it must divide either b or q; by the same token, if a prime divides the numerator a and does not divide the denominator b, it must divide the quotient q. In our present circumstance, this means the quotient $\binom{p}{k}$ must be divisible by p for $k = 1, 2, \ldots, p - 1$, so that mod p the binomial theorem reads

$$2^p \equiv 2 \pmod{p}.$$

Almost *none* of this argument works if p is replaced by a nonprime number n. At least the binomial theorem

$$2^n = (1+1)^n = 1 + \binom{n}{1} + \binom{n}{2} + \cdots + \binom{n}{n-1} + 1.$$

still holds. But when we ask which binomial coefficients $\binom{n}{k}$ are divisible by n, we're lost. We still have the formula

$$\binom{n}{k} = \frac{n(n-1)(n-2)\ldots(n-k+1)}{k(k-1)(k-2)\ldots 1}.$$

but it doesn't tell us much. For one thing, if n is not prime, it's possible for a product of numbers to be divisible by n even if none of the factors are; thus, even though the terms in the denominator are each smaller than n, their product may well be divisible by n.

What's more, even if the denominator were not divisible by n and the numerator were, it wouldn't follow that the quotient was divisible by n: after all, 6 divides the numerator of the fraction $12/3$ and not the denominator, but it doesn't divide the quotient.

In fact, it's very difficult to say when a binomial coefficient $\binom{n}{k}$ is divisible by n, except when n is a prime. This approach to proving Fermat's Theorem completely fails in the case of a nonprime modulus.

20.3.2 The Proof that Succeeds

Let's consider now the other proof of Fermat's Theorem, to see if it does any better. That proof was based on the observation that, in the multiplication table for arithmetic mod p, every number appears once in each (nonzero) row: in other words, for any nonzero number a, the sequence of numbers

$$a, 2a, 3a, \ldots, (p-2)a, (p-1)a$$

is just a reshuffling of the sequence

$$1, 2, 3, \ldots, p-2, p-1.$$

The reason for this is straightforward: every nonzero number a has a reciprocal b in arithmetic mod p, that is, a number such that $ab \equiv 1 \pmod{p}$. This means any number c can be written as

$$c \equiv (ab) \cdot c \equiv a \cdot (bc);$$

in other words, every number is a multiple of a in arithmetic mod p.

Now, given that the two sequences above are the same except for order, we look at various ways of writing the number a in arithmetic mod p:

$$a \equiv \frac{a}{1} \equiv \frac{2a}{2} \equiv \frac{3a}{3} \equiv \cdots \equiv \frac{(p-2)a}{p-2} \equiv \frac{(p-1)a}{p-1} \pmod{p}.$$

If we multiply together these $p-1$ different expressions for a, we see that

$$a^{p-1} \equiv \frac{a \cdot 2a \cdot 3a \cdot \cdots \cdot (p-2)a \cdot (p-1)a}{1 \cdot 2 \cdot 3 \cdot \cdots \cdot (p-2) \cdot (p-1)} \pmod{p}$$

and since the numerator is just a reordering of the denominator and both are nonzero, we conclude that

$$a^{p-1} \equiv 1 \pmod{p}.$$

Does this make sense if we replace p by a composite number n, and assume that a is relatively prime to n? It's still true in this case that a will have a reciprocal mod n, and therefore the sequences

$$a, 2a, 3a, \ldots, (n-2)a, (n-1)a$$

and

$$1, 2, 3, \ldots, n-2, n-1.$$

are just reshufflings of each other. But the argument goes out the window at the next step: if, for example, n is even, we can't write

$$a \equiv \frac{2a}{2} \pmod{n}.$$

because *when n is even, we can't divide by 2 in arithmetic mod n.*

Perhaps you're thinking: "What harm can there be in canceling 2s in an expression like $\frac{2a}{2}$?" Here's what goes wrong. Suppose we take n to be 10, and a to be 3. It's still true that the nonzero multiples of 3

$$3, 6, 9, 2, 5, 8, 1, 4, 7$$

in arithmetic mod 10 are the same as the nonzero numbers

$$1, 2, 3, 4, 5, 6, 7, 8, 9$$

in another order. We could then write

$$3 \equiv \frac{3}{1} \equiv \frac{6}{2} \equiv \frac{9}{3} \equiv \frac{2}{4} \equiv \frac{5}{5} \equiv \frac{8}{6} \equiv \frac{1}{7} \equiv \frac{4}{8} \equiv \frac{7}{9} \pmod{10}$$

and then multiply these nine fractions together to get

$$3^9 \equiv \frac{3 \cdot 6 \cdot 9 \cdot 2 \cdot 5 \cdot 8 \cdot 1 \cdot 4 \cdot 7}{1 \cdot 2 \cdot 3 \cdot 4 \cdot 5 \cdot 6 \cdot 7 \cdot 8 \cdot 9} \pmod{10}$$

We could then rearrange the terms in the numerator to conclude that

$$3^9 \equiv 1 \pmod{10}$$

What's wrong with that?

The answer is, both the numerator and the denominator of the fraction in that last expression are equal to 0 (mod 10), which renders the expression meaningless. More to the point: the result is just wrong! As you can check:

$$3^9 \equiv 3^4 \cdot 3^4 \cdot 3 \equiv 81 \cdot 81 \cdot 3 \equiv 3 \pmod{10}.$$

We've ignored the dire warning of Section 17.5—we've tried to divide in arithmetic mod n by a number not relatively prime to n— and this is the result.

Can we salvage anything from this? We can, and in fact cleaning up this mess will lead us to the correct statement and proof of Euler's Theorem. The fix is simplicity itself: in the series of expressions

$$a \equiv \frac{a}{1} \equiv \frac{2a}{2} \equiv \frac{3a}{3} \equiv \cdots \equiv \frac{(n-2)a}{n-2} \equiv \frac{(n-1)a}{n-1} \pmod{n}$$

we have to use only the legitimate equalities: the ones whose denominators are relatively prime to n. For example, in the case we just considered—$n = 10$ and $a = 3$—we only look at the four fractions whose denominators are 1, 3, 7 and 9. That is, we write

$$3 \equiv \frac{3}{1} \equiv \frac{9}{3} \equiv \frac{1}{7} \equiv \frac{7}{9} \pmod{10}$$

and then we multiply together these four expressions for 3 in arithmetic mod 10 to see that

$$\begin{aligned} 3^4 &\equiv \frac{3 \cdot 9 \cdot 1 \cdot 7}{1 \cdot 3 \cdot 7 \cdot 9} \\ &\equiv 1 \pmod{10} \end{aligned}$$

which is not only legitimate, it's true! What makes this work is that, when we multiply by 3 in arithmetic mod 10, numbers relatively prime to 10 are carried into other numbers relatively prime to 10, and vice versa. In other words, if we write out the sequence of numbers relatively prime to 10

$$1, 3, 7, 9$$

and then multiply by 3,

$$3, 9, 1, 7$$

we again see the same sequence in a different order.

Let's see how this goes for an arbitrary modulus n, and a number a relatively prime to n. Before we begin, though, recall the definition of the Euler ϕ-function: $\phi(n)$ is simply the number of numbers between 0 and $n-1$ that are relatively prime to a given number n. If you're a little rusty on this material, you might want to review the discussion in Chapter 13 and in particular the recipe given in Section 13.2 for evaluating $\phi(n)$.

Ready? We'll illustrate the general method, in steps, with an example, $n = 21$ and $a = 2$.

- Write out the numbers between 1 and $n-1$ that are relatively prime to n. We know how many there should be: this is just the Euler function $\phi(n)$, which we saw how to evaluate earlier. In our example, $n = 21 = 3 \times 7$, so $\phi(n) = 2 \times 6 = 12$; the 12 numbers between 0 and 20 that are relatively prime to 21 are

$$1, 2, 4, 5, 8, 10, 11, 13, 16, 17, 19, 20$$

- Now multiply each of these numbers by a. Since a is relatively prime to n, multiplying a number relatively prime to n by a will yield another number relatively prime to n, so the sequence you see here should be the same as the first, but in a different order. This checks out in our example: multiplying the numbers above by 2 in arithmetic mod 21, we get

$$2, 4, 8, 10, 16, 20, 1, 5, 11, 13, 17, 19$$

respectively.

- The number a will now be equal to each number on your second list divided by the corresponding number of your first list: in our example,

$$\begin{aligned} 2 &\equiv \frac{2}{1} \equiv \frac{4}{2} \equiv \frac{8}{4} \equiv \frac{10}{5} \equiv \frac{16}{8} \equiv \frac{20}{10} \equiv \frac{1}{11} \equiv \frac{5}{13} \equiv \frac{11}{16} \\ &\equiv \frac{13}{17} \equiv \frac{17}{19} \equiv \frac{19}{20} \pmod{21}. \end{aligned}$$

- You have now a total of $\phi(n)$ fractions, each of which equals a in arithmetic mod n. Finally, multiply them together. Observe that since every number on your list—in other words, every number between 1 and $n - 1$ relatively prime to n—appears exactly once in the numerator and once in the denominator, the quotient is exactly 1. In our example, we see that

$$\begin{aligned} 2^{12} &\equiv \frac{24 \cdot 8 \cdot 10 \cdot 16 \cdot 20 \cdot 1 \cdot 5 \cdot 11 \cdot 13 \cdot 17 \cdot 19}{1 \cdot 2 \cdot 4 \cdot 5 \cdot 8 \cdot 10 \cdot 11 \cdot 13 \cdot 17 \cdot 19 \cdot 20} \\ &\equiv 1 \pmod{21}. \end{aligned}$$

In general, we may deduce that

$$a^{\phi(n)} \equiv 1 \pmod{n}.$$

Of course, you don't need to go through these steps each time: the point is simply that you could, and that the end result would be as stated. That's what a proof is, and the result we've established is what's called *Euler's Theorem*:

> For any number n and any number a relatively prime to n,
>
> $$a^{\phi(n)} \equiv 1 \pmod{n}$$
>
> where ϕ is the Euler function.

Note that this is a direct extension of Fermat's Theorem: if n happens to be equal to a prime number p, then every number between 1 and $p - 1$ is relatively prime to p, so $\phi(p) = p - 1$. In this case, Euler's Theorem is the earlier result claimed by Fermat.

20.4 Using Euler's Theorem

As you might expect, Euler's Theorem allows us to do, in arithmetic modulo any number n, what Fermat's Theorem allowed us to do modulo a prime number p. The two main applications are to finding powers of a number mod n, and finding roots mod n; we'll discuss these in turn. In each case the process is not fundamentally different from the case of prime modulus.

We'll start with finding powers. Recall how we used Fermat's Theorem to do this modulo a prime: for example, when we were asked to calculate $11^{56} \pmod{19}$, we invoked Fermat's Theorem, which told us that

$$11^{18} \equiv 1 \pmod{19}.$$

Moreover, the same thing is true for 11^{36}, 11^{54} and so on, because

$$11^{36} \equiv (11^{18})^2 \equiv 1 \pmod{19},$$
$$11^{54} \equiv (11^{18})^3 \equiv 1 \pmod{19},$$

Hence, we can write

$$11^{56} \equiv 11^{54} \cdot 11^2 \equiv 11^2 \equiv 121 \equiv 7 \pmod{19}.$$

We can do exactly the same thing to evaluate—or at least reduce—a power a^k of a number a in arithmetic mod n for any modulus n: by using Euler's Theorem, we can always reduce the exponent k to a number less than $\phi(n)$. Explicitly, what we do is

- First, we find $\phi(n)$. Recall how we do this: start with n, and for each prime p dividing n multiply by $\frac{p-1}{p}$.
- Second, we write the exponent k as a multiple of $\phi(n)$, plus a remainder $r < \phi(n)$:

$$k = m \cdot \phi(n) + r.$$

- Next, we write

$$a^k \equiv (a^{\phi(n)})^m \cdot a^r$$
$$\equiv a^r \pmod{n}.$$

- Finally, we evaluate $a^r \pmod{n}$ to arrive at the answer.

Of course, if $\phi(n)$ is large, r may be too, and we may have to work to evaluate a^r; but at least we've reduced the problem: we know that r is a number less than $\phi(n)$. Note also that while this trick may save us time, it's not necessary: we can always evaluate powers by the method of Section 18.2.

Here's an example:

EXAMPLE 20.4.1 Evaluate $6^{441} \pmod{65}$.

SOLUTION The first step is to find $\phi(65)$. Since 65 factors as 5×13, we have

$$\phi(65) = 65 \cdot \frac{4}{5} \cdot \frac{12}{13}$$
$$= 4 \cdot 12$$
$$= 48.$$

Thus, Euler's Theorem tells us that $a^{48} \equiv 1 \pmod{65}$ for any number a relatively prime to 65. Since 6 is relatively prime to 65, this applies.

Next, we divide $\phi(65) = 48$ into the exponent 441: in other words, we write

$$441 = 9 \times 48 + 9.$$

Then Euler's Theorem tells us that

$$6^{441} \equiv (6^{48})^9 \cdot 6^9$$
$$\equiv 6^9 \pmod{65}.$$

Finally we have to evaluate $6^9 \pmod{65}$, which we do by the method of Section 18.2: we write the exponent 9 as a sum of powers of 2

$$9 = 8 + 1$$

and make a table of powers of 6 (mod 65) by successive squaring:

$$6^1 \equiv 6 \pmod{65}$$
$$6^2 \equiv 36 \pmod{65}$$
$$6^4 \equiv 1{,}296 \equiv -4 \pmod{65}$$
$$6^8 \equiv 16 \pmod{65}.$$

We have then

$$6^{441} \equiv 6^9 \equiv 6 \cdot 16 \equiv 96 \equiv 31 \pmod{65}.$$ ■

Our second application of Euler's Theorem will be to finding roots in arithmetic mod n. Again, this is based on the same idea as the method we worked out in Section 19.4 when the modulus is prime, but it can be more complicated, so it's worth reviewing what we did there.

The basic idea of Section 19.4, if you recall, was to use the cyclical nature of powers in arithmetic mod p to express a root of a number a as a power of $a \pmod{p}$. The idea was that, as long as a was not divisible by p,

$$a^{p-1} \equiv 1 \pmod{p}.$$

Therefore we see that

$$a \equiv a^p \equiv a^{2p-1} \equiv a^{3p-2} \equiv \dots \pmod{p}.$$

Now, if we wanted to find the kth root $\sqrt[k]{a}$ of a in arithmetic mod p, we just have to find one of these expressions for a that is a kth power: that is, whose exponent is divisible by k. In other words, we find a number of the form $\ell(p-1)+1$ that is simultaneously a multiple mk of k. This amounts to solving the equation

$$mk = \ell(p-1)+1,$$

as we saw how to do by the Euclidean algorithm in Section 9.3; or, equivalently, finding the reciprocal m of k in arithmetic mod $p-1$, which we saw in Section 17.3 how to do—as long as k is relatively prime to $p-1$. Having done this, we write

$$a \equiv a^{\ell(p-1)+1} \equiv (a^m)^k \pmod{p}.$$

We see then that a^m is a kth root of a in arithmetic mod p.

We are now in pretty much the same situation modulo any number n. Euler's Theorem tells us that the powers of any number a relatively prime to n go in cycles of length $\phi(n)$, so that we have

$$a \equiv a^{\phi(n)+1} \equiv a^{2\phi(n)+1} \equiv a^{3\phi(n)+1} \equiv \dots \pmod{n}$$

Again, the game is to find one of these exponents $\ell\phi(n)+1$ that is simultaneously a multiple mk of k. If we can do that, we can write

$$a \equiv a^{\ell\phi(n)+1} \equiv (a^m)^k \pmod{n}.$$

and again the power a^m will be the kth root of $a \pmod p$. The moral, simply stated, is this: because the powers of a go in cycles of length $\phi(n)$, *if m and k are reciprocals* $(\text{mod } \phi(n))$, *then the* k^{th} *root and the* m^{th} *power are the same* $(\text{mod } n)$.

Let's translate this into a procedure, or algorithm, which we'll then illustrate with an example.

How to Find $\sqrt[k]{a} \pmod n$

- Find $\phi(n)$. Without this, we can't proceed.
- Check that a is relatively prime to n, and that k is relatively prime to $\phi(n)$. Both of these are absolutely necessary for the method to work: if a isn't relatively prime to n then Euler's Theorem won't apply; and if k isn't relatively prime to $\phi(n)$ we won't be able to find a reciprocal of k $(\text{mod } \phi(n))$. Indeed, if either of these conditions doesn't hold, there may not be a kth root of $a \pmod n$, or there may be more than one k^{th} root.
- Find the reciprocal m of k in arithmetic mod $\phi(n)$: that is, use the Euclidean algorithm to solve the equation

$$mk = \ell \cdot \phi(n) + 1.$$

- Find the mth power of $a \pmod n$; since

$$(a^m)^k \equiv a^{mk} \equiv a^{\ell \cdot \phi(n)+1} \equiv a \pmod n,$$

this is the answer.

- Check the answer by raising it to the kth power and seeing that you do indeed get $a \pmod n$. (This isn't officially part of the process, but if you think you can consistently get through all the above steps without making a mistake, you have more confidence in your computational skills than we have in ours.)

Got all that? If it seems a little bewildering at first, don't worry: this is genuinely a complex procedure. In particular, keep in mind that while the basic problem is an equation in modular arithmetic mod n, the third step involves a calculation in modular arithmetic mod $\phi(n)$, so that you have to switch between these two number systems. This is not a serious problem, as long as you keep your wits about you, but it may take some practice to gain a facility with the process.

Speaking of practice, let's do one concrete problem to illustrate the idea.

EXAMPLE 20.4.2 In arithmetic mod 85, find a fifth root of 3—that is, find a number x such that

$$x^5 \equiv 3 \pmod{85}.$$

SOLUTION We'll go step by step.

Step 1. Find $\phi(85)$. This is straightforward: $85 = 5 \times 17$, so

$$\phi(85) = 85 \cdot \frac{4}{5} \cdot \frac{16}{17} = 64.$$

Step 2. Check that 3 is relatively prime to the modulus 85 (it is), so that we can apply Euler's Theorem to powers of $3 \pmod{85}$; and that 5 is relatively prime to $\phi(85) = 64$ (it is), so that we can find a reciprocal of $5 \pmod{64}$.

Step 3. Find the reciprocal of 5 (mod 64). This is also easy: $5 \times 13 = 65 \equiv 1$ (mod 64), so 13 is the reciprocal of 5 (mod 64). (If we didn't see the answer here right away, we could do it by the Euclidean algorithm.) Since $3^{64} \equiv 1 \pmod{85}$, this means that

$$\begin{aligned}(3^{13})^5 &= 3^{65} \\ &= 3^{64+1} \\ &\equiv 3 \pmod{85}\end{aligned}$$

so $x = 3^{13}$ will be our answer.

Step 4. Evaluate 3^{13} (mod 85). Also straightforward. We start by writing 13 as a sum of powers of 2:

$$13 = 8 + 4 + 1$$

and then make a table of powers of 3 (mod 85) by successive squaring:

$$\begin{aligned}3^2 &\equiv 9 \pmod{85} \\ 3^4 &\equiv 9^2 \equiv 81 \equiv -4 \pmod{85} \\ 3^8 &\equiv (-4)^2 \equiv 16 \pmod{85}\end{aligned}$$

so

$$\begin{aligned}3^{13} &\equiv 3^8 \times 3^4 \times 3 \pmod{85} \\ &\equiv 16 \times (-4) \times 3 \pmod{85} \\ &\equiv 63 \pmod{85}\end{aligned}$$

and 63 is the answer (we think).

Step 5. Check that 63^5 really is 3 (mod 85). We'll leave this for you to do. Let us know if we're right. ■

As we said, this process can be confusing, as it involves calculations in two different number systems: mod n and mod $\phi(n)$. Still, with a little practice you should get the hang of it. Try this.

Exercise 20.4.3 For each of the following roots in arithmetic mod n, say whether or not the method described here applies; and if it does find the root.

1. $\sqrt[5]{6}$ in arithmetic mod 21
2. $\sqrt[3]{5}$ in arithmetic mod 15
3. $\sqrt[5]{2}$ in arithmetic mod 21
4. $\sqrt[3]{3}$ in arithmetic mod 28
5. $\sqrt[5]{6}$ in arithmetic mod 35

20.5 Whose Idea Was This, Anyway?

Before we close this part of the book, we'd like to talk a little about how the ideas we've seen here were developed. Who first thought of modular arithmetic, and who first discovered its properties?

Actually, there is no simple answer to that question. Fermat, for example, was the one who first stated the theorem that bears his name, in the 17th century; but he didn't express it in the language of modular arithmetic. Rather, he stated it as a divisibility property: in slightly modernized language, what he said was, "Suppose p is a prime number, and a any number not divisible by p. If you raise the number a to the $(p-1)^{\text{st}}$ power and subtract 1, the result is divisible by p."

These ideas were extended by Leonhard Euler in the 18th century. Euler encountered Fermat's work in his twenties, and tried to find proofs of the statements therein for much of his life. (He did many other things as well—his published works run to seventy volumes!) He went on to introduce the ϕ-function and state and prove his extension of Fermat's Theorem to arithmetic mod n. But for the most part, he stated his results, as did Fermat, in terms of divisibility.

It was with the work of Carl Friedrich Gauss in the beginning of the 19th century that modular arithmetic really came into its own. Gauss invented the modern notation for arithmetic mod n, and proved some of the great theorems in number theory, in his treatise *Disquisitiones Arithmeticae*. (This book, by the way, appeared in 1801, when he was twenty-four years old! We reproduce two pages from it on pgs. 229–230.) Gauss was a man of awesome mental talents: he invented both the mathematical subject of Differential Geometry and the modern theory of geodesy while surveying the province of Brunswick for his employer.

It's amazing how much difference a change of notation can make. Gauss' work certainly went beyond what Euler had done, but it was his introduction of the modern notation for modular arithmetic that really freed succeeding generations of mathematicians to think about the subject in the way we do now. It was this notation that led, later in the century, to the viewpoint on modular arithmetic that we've taken in this book—modular arithmetic as an alternative number system—and indeed some might say that it was this example that led to the whole notion of a "number system" developed in the latter 19th and early 20th century. It's truly the case that the soul of mathematics often lies not in its theorems, but in its definitions and notation.

— 2 —

vero ad fractos sunt extendendae. *E. g.* -9 et $+16$ secundum modulum 5 sunt congrui; -7 ipsius $+15$ secundum modulum 11 residuum, secundum modulum 3 vero nonresiduum. Ceterum quoniam cifram numerus quisque metitur, omnis numerus tamquam sibi ipsi congruus secundum modulum quemcunque est spectandus.

2. Omnia numeri dati a residua secundum modulum m sub formula $a + km$ comprehenduntur, designante k numerum integrum indeterminatum. Propositionum quas post trademus faciliores nullo negotio hinc demonstrari possunt: sed istarum quidem veritatem aeque facile quiuis intuendo poterit perspicere.

Numerorum congruentiam hoc signo, $\equiv$, in posterum denotabimus, modulum vbi opus erit in clausulis adiungentes, $-16 \equiv 9$ (mod. 5), $-7 \equiv 15$ (mod. 11) *).

3. Theor. *Propositis m numeris integris successiuis, a, $a+1$, $a+2 \ldots\ldots a+m$ 1, alioque A, illorum aliquis huic secundum modulum m congruus erit, et quidem vnicus tantum.*

Si enim $\frac{A-a}{m}$ integer, erit $a \equiv A$, sin fractus, sit integer proxime maior, (aut quando est negatiuus, proxime *minor*, si ad signum non respiciatur) $= k$, cadetque $A + km$ inter a et

*) Hoc signum propter magnam analogiam quae inter aequalitatem atque congruentiam inuenitur adoptauimus. Ob eandem caufsam ill. Le Gendre in comment infra saepius laudanda ipsum aequalitatis signum pro congruentia retinuit, quod nos ne ambiguitas oriatur imitari dubitauimus.

The page of Gauss' *Disquisitiones* where the notation for congruence is introduced.

ex his 1, 2, 3 ... $p-1$, quae omnes tum inter se quam a numeris in (A) et (B) contentis erunt diuersi. Assertiones priores eodem modo demonstrantur vt in II, tertia ita. Si esset $\gamma a^m \equiv \beta a^n$, fieret $\gamma \equiv \beta a^{n-m}$, aut $\equiv \beta a^{t+n-m}$ prout $m < n$, aut $> n$, in vtroque casu γ alicui ex (B) congrua contra hyp. Habentur igitur $3t$ numeri ex his 1, 2, 3 ... $p-1$, atque si nulli amplius desunt, fiet $t = \frac{p-1}{3}$ adeoque theorema erit demonstratum.

IV. Si vero etiamnum aliqui desunt eodem modo ad quartum numerorum complexum, (D), progrediendum erit etc. Patet vero quoniam numerorum 1, 2, 3.... $p-1$ multitudo est finita, tandem eam exhaustum iri, adeoque multiplum ipsius t fore: quare t erit pars aliquota numeri $p-1$. *Q. E. D.*

50. Quum igitur $\frac{p-1}{t}$ sit integer, sequitur euehendo vtramque partem congruentiae $a^t \equiv 1$ ad potestatem exponentis $\frac{p-1}{t}$, $a^{p-1} \equiv 1$, *siue* $a^{p-1} - 1$ *semper per p diuisibilis est, quando p est primus ipsum a non metiens.*

Theorema hoc quod tum propter elegantiam tum propter eximiam vtilitatem omni attentione dignum, ab inuentore *theorema Fermatianum* appellari solet. Vid. *Fermatii Opera Mathem. Tolosae* 1679 *fol. p.* 163. Demonstrationem inuentor non adiecit, quam tamen in potestate sua esse professus est. Ill. Euler primus demonstrationem publici iuris fecit, in diss. cui titulus *Theorematum quorundam ad numeros primos spectantium demonstratio*, *Comm. Acad. Petrop.* T.

Gauss' discussion of Fermat's Theorem, from the *Disquisitiones*.

"Two added to one—if that could but be done,"
It said, "with one's fingers and thumbs!"

Recollecting with tears how, in earlier years,
It had taken no pains with its sums.

"The thing can be done," said the Butcher, "I think.
The thing must be done, I am sure.

The thing shall be done! Bring me paper and ink,
The best there is time to procure."

The Beaver brought paper, portfolio, pens,
And ink in unfailing supplies:

While strange creepy creatures came out of their dens,
And watched them with wondering eyes.

So engrossed was the Butcher, he heeded them not,
As he wrote with a pen in each hand,

And explained all the while in a popular style,
Which the Beaver could well understand.

"Taking Three as the subject to reason about—
A convenient number to state—

We add Seven, and Ten, and then multiply out
By One Thousand diminished by Eight.

"The result we proceed to divide, as you see,
By Nine Hundred and Ninety and Two:

Then subtract Seventeen, and the answer must be
Exactly and perfectly true.

"The method employed I would gladly explain,
While I have it so clear in my head,

If I had but the time and you had but the brain—
But much yet remains to be said.

"In one moment I've seen what has hitherto been
Enveloped in absolute mystery,

And without extra charge I will give you at large
A Lesson in Natural History."

Lewis Carroll, *in The Hunting of the Snark*

PART IV

Codes and Primes

You Deserve a Break

Are you still with us? If so, give yourself a pat on the back: if you look back over the preceding part of the book, you'll see we've come quite a long way from the days of introducing modular arithmetic. There's much more to say about the wonderful, peculiar world of modular arithmetic—there are mathematicians who spend their lives investigating it—but at this point you know enough to count yourself among the initiates. After all, few enough people even know what it means to find the fifth root of 3 in arithmetic mod 85, and fewer still know how to do it; you do. And you certainly know more than enough to develop some interesting applications.

In any event, if you've gotten to this point, you deserve a break. And in fact you get one—two, actually. Specifically:

- ***It gets easier from here on out.*** We still have some things to do that are going to require thought, but we're not going to ask you, in the remainder of this book, to master anything else as demanding as the techniques of the last few sections. In other words, if you've come this far you can finish: it'll just take a little stamina. Plus,
- ***It gets more real.*** When we embarked, a hundred pages or so ago, on our excursion into the world of modular arithmetic, we did so in a spirit of pure intellectual curiosity. Certainly we wouldn't use modular arithmetic to model real-world situations, and so we wouldn't expect it to have any real-world applications. But we have a remarkable story to tell in the last part of the book: how it was discovered that the particular properties of modular arithmetic could be used to generate a new kind of code, one that has revolutionized cryptography and made electronic commerce possible. What's more, in order to set up these codes, and to see why they work the way they do, we need only what we've already learned about modular arithmetic: specifically, the techniques we've described for taking powers and roots in arithmetic mod n.

So here's what we're going to do in the last part of this book. We're going to start by talking about codes in general, and what makes a good code versus a bad one. Then, in the second chapter, we'll show how modular arithmetic was used to create an entirely new kind of code, called a *public-key cryptosystem*.

One thing we'll see about public key cryptography is that we need to have large prime numbers to make it work. How can we find large primes? This just leads to a further problem: how can we tell when a particular number is prime? We'll give the answer, and wrap up the discussion of finding large primes, in Chapter 23.

All this leads us back, in the final chapter, to two of the central themes of this book: the structure of arithmetic mod p; and the perennially fascinating topic of how primes are distributed among all numbers. Specifically, we'll talk about the notion of a *generator* for arithmetic mod p (and the very cute application of this notion to verifying electronic signatures), and the related question of congruence classes of primes. It seems fitting to end the book here, with an investigation that weaves together many of the ideas we've encountered along the way.

Chapter 21

Codes

Eons ago, in an epoch long since shrouded in the mists of forgotten history, humankind first conceived the idea of written language. Probably around a week or so later, two people felt the need to send messages to each other without a third party being able to understand them, and the earliest codes were born.

In this chapter, we'll take you through some of the history of codes since then, describing some of the classic methods that have been developed for encoding and decoding messages. Mostly we're doing this because it's historically fascinating (and fun), but it also serves a mathematical purpose: by the end of the chapter, we'll be able to understand a fundamental defect of all the codes described here, and how the coding system we'll introduce in the following chapter manages to avoid it.

21.1 Substitution Codes

This is where coding starts for most of us. The idea is simple: we just use a different alphabet from the standard, substituting the symbols of our alphabet for the standard ones according to a key. The simplest of these—everybody's done this once when they were kids—is to use numbers in place of letters, replacing a letter by its position in the alphabet: 1 for A, 2 for B, and so on. The key would thus be

A	B	C	D	E	F	G	H	I	J	K	$L\ldots$
1	2	3	4	5	6	7	8	9	10	11	12 ...

and so on; the message

this is a dumb code

would be encrypted as

20-8-9-19 9-19 1 4-21-13-2 3-15-4-5

Of course, we don't have to use numbers in place of letters; we could, for example, use other letters. One popular code (in elementary school, at least) is simply to shift every letter by one in the alphabet cyclically: use B for A, C for B,

D for C and so on up to Z for Y and A for Z. The key would be

A	*B*	*C*	*D*	*E*	*F*	*G*	*H*	*I*	*J*	*K*	*L* ...
B	*C*	*D*	*E*	*F*	*G*	*H*	*I*	*J*	*K*	*L*	*M* ...

and so on; the message

this is even worse

would be encrypted as

UIJT JT FWFO XPSTF

As a variant on this, we could shift every letter by any fixed amount, for example, by seven letters:

A	*B*	*C*	*D*	*E*	*F*	*G*	*H*	*I*	*J*	*K*	*L* ...
H	*I*	*J*	*K*	*L*	*M*	*N*	*O*	*P*	*Q*	*R*	*S* ...

with S replaced by Z, T replaced by A, U by B and so on. In other words, if we associate to each letter a number between 1 and 26 according to its position in the alphabet, and vice versa, this code consists simply of adding 7 in arithmetic mod 26.

Now, all of these codes involve a systematic substitution. This makes it easy to remember what the key is, and to communicate it to your co-conspirator. But it also makes it really easy to crack the code: once someone guesses what letter a few of the symbols stand for they can figure out the rest. It's more effective to choose symbols (letters, numbers or others; it doesn't matter) at random. For example, we could choose numbers and typographic symbols as follows:

A	*B*	*C*	*D*	*E*	*F*	*G*	*H*	*I*	*J*	*K*	*L*	*M*
5	2	–	†	8	1	3	4	6	%	{	0	9
N	*O*	*P*	*Q*	*R*	*S*	*T*	*U*	*V*	*W*	*X*	*Y*	*Z*
*	‡	.	+	(	)	;	?	¶	§	#	;	}

By way of example, here's a message encrypted according to this last code: in fact, it's the coded message that appears in the story "*The Gold-Bug*" by Edgar Allan Poe, and whose unraveling you'll read about later on in this chapter.

```
53‡‡†305))6*;4826)4‡.)4‡;806*;48†8¶60))85;1‡(;:‡*8†83(88)
5*†;46(;88*96*?;8)*‡(;485);5*†2:*‡(;4956*2(5*_4)8¶8*;40692
85);)6†8)4‡‡;1(‡9;48081;8:8‡1;48†85;4)485†528806*81(‡9;48;
(88;4(‡?34;48)4‡;161;:188;‡?;
```

Now, if the message we wanted to encode were short enough—a word or two, say—there would be no way for a third party to decode the message without knowing the key. But for messages of any substantial length—the one immediately above, for example—there are a number of techniques for cracking the code. We'll talk a little about those here.

1. ***Letter frequency.*** As everyone knows (especially anyone who's played Scrabble), not all letters occur equally often in English. The most common is *e* (in a typical text, 12.7% of the letters are *e*) followed in order by *t* (9.1%), *a* (8.2%), *o* (7.5%), *n* (6.7%) and *s* (6.3%). At the opposite end of the spectrum, *q* and *z* make up only 0.1% of the letters in an average text, while *j* and *x* account for 0.2% each.

Of course, any given text will deviate from these percentages. The shorter it is, the less likely it is to conform to this pattern: a two or three word text may not involve any *e*'s at all. But when we look at longer texts the percentages do conform very closely with these figures, unless the author has made a deliberate attempt to avoid one letter or overuse another.[1] Thus, for example, in the coded message above from "*The Gold-Bug*," it is likely that the most commonly occurring symbol represents an *e*; and while the message is not long enough to conclude with any certainty that the second most frequent letter represents *t*, if it were much longer we would probably be justified in making that assumption.

2. *Juxtapositions.* Some letters in English frequently appear doubled: among the vowels, for example, *ee* and *oo* are common, while *aa* and *ii* are much less so. Among consonants, likewise, *c* is more common than *f*, *m* or *p*; but the double letters *ff*, *mm* and *pp* all occur much more frequently than *cc*. Thus, seeing what fraction of the occurrences of a given symbol are doubles gives us clues about the identity of that symbol.

It's more generally the case that some letters tend to come after others. The most extreme example is of course that *u* usually follows *q*, but there are many others: *ck* is common, for example, while *kc* is relatively rare. In fact, we can make a distinction among letters: some, like *a*, are *democratic*, meaning that they come regularly both before and after all other letters; while others—*t*, for example—most often come after vowels or consonants like *n*, *r* or *s*, and much less commonly after other consonants (for example, *d*, *g*, *k* or *m*). Thus if we were trying to determine whether a symbol $\diamond$ in an encoded text represented *a* or *t*, and we couldn't say on the basis of frequency, we could tabulate the symbols occurring immediately before $\diamond$. If all other symbols occurred before $\diamond$ with more or less the same frequency that they did in the text as a whole—if the symbol $\diamond$ was democratic, in other words—it would more likely represent *a*; if $\diamond$ was less democratic, it would probably represent *t*.

Again, no document is apt to conform exactly to the average percentages of English texts in general; and it's especially tricky to draw inferences from juxtapositions of letters in a text where spaces between words are not indicated (as in the case of the message above, for example). But once more, the longer the message the more information we can glean in this way.

3. *Repeated sequences.* There are some sequences of letters that occur very frequently in English: the sequence "the," for example. It's so common, in fact, that one of the first things a codebreaker will do when faced with an encoded message is to search for frequently-occurring triples; it's a good bet that sequences like "the" and "and" will be among them.

4. *Trial and error.* (Otherwise known as the thousand-monkey method.) Suppose we find ourselves with a coded message, involving a total of 26 different symbols.

---

[1] In fact Georges Perec wrote a full-length novel in French, *La disparition*, in which no word containing an *e* is used. (*e* is also the most common letter in French.) What's more, it was then translated into English—still without using the letter *e*. We have no idea why.

We suspect that it's been encoded using a straight substitution code, so we simply have to figure out how the 26 symbols correspond to the 26 letters of the ordinary English alphabet.

Well, how many possible ways could they correspond? In other words, how many ways are there of assigning one of the letters of the alphabet to each of 26 symbols? If you managed to stay awake during the first part of this book, you know the answer: it's 26 factorial, or 26!, which is equal to

$$403{,}291{,}461{,}126{,}605{,}635{,}584{,}000{,}000.$$

Now, for a single codebreaker working by hand, that probably puts it out of reach of practicality: trying each possible assignment until you find one that works would take roughly the lifetime of the universe. But modern computing technology has made feasible a lot of processes that once were prohibitively time-consuming. It's still true that even with a supercomputer searching through all 26! permutations of the letters $A, B, \ldots, Z$ would still take more time than we have on this earth. But if we can narrow down the list of possible permutations at all significantly we may be in range.

This of course presumes that the message, once decoded, will make sense—in other words, that we'll recognize when we've succeeded in decoding it. If the message consisted of a meaningless string of letters or numbers—a credit card number, for example—we'd have no way of knowing when we got it right.

21.2 The Gold-Bug

At this point it might be fun to see how some of these methods might be carried out in practice. Rather than make up an example ourselves, we thought it would be more enjoyable to read the relevant part of the classic short story *The Gold-Bug* by Edgar Allan Poe, in which decryption plays a part.

At the start of the segment quoted here, the narrator and his host Legrand have just returned from a successful hunt for buried treasure. The narrator is completely mystified as to how Legrand knew where to look. Legrand tells of finding an antique parchment, on which he discerns drawings of a death's-head and a goat—references, he thinks, to the pirate Captain Kidd. The passage opens with the narrator urging Legrand to continue with his tale ...

"But proceed—I am all impatience."

"Well; you have heard, of course, the many stories current—the thousand vague rumors afloat about money buried, somewhere on the Atlantic coast, by Kidd and his associates. These rumors must have had some foundation in fact. And that the rumors have existed so long and so continuously could have resulted, it appeared to me, only from the circumstance of the buried treasure still *remaining* entombed. Had Kidd concealed his plunder for a time, and afterwards reclaimed it, the rumors would scarcely have reached us in their present unvarying form. You will observe that the stories told are all about money-seekers, not about moneyfinders. Had the pirate recovered his money; there the affair would have dropped. It seemed to me that some accident—say the loss of a memorandum indicating its locality—had deprived him of the means of recovering it, and that this accident had become known to his followers, who otherwise might never have heard that treasure had been concealed at all, and who, busying themselves in vain, because unguided attempts to regain it, had given first birth, and then universal currency, to the reports which are now so common. Have you ever heard of any important treasure being unearthed along the coast?"

"Never."

"But that Kidd's accumulations were immense, is well known. I took it for granted, therefore, that the earth still held them; and you will scarcely be surprised when I tell you that I felt a hope, nearly amounting to certainty, that the parchment so strangely found, involved a lost record of the place of deposit."

"But how did you proceed?"

"I held the vellum again to the fire, after increasing the heat; but nothing appeared. I now thought it possible that the coating of dirt might have something to do with the failure; so I carefully rinsed the parchment by pouring warm water over it, and, having done this, I placed it in a tin pan, with the skull downwards, and put the pan upon a furnace of lighted charcoal. In a few minutes, the pan having become thoroughly heated, I removed the slip, and, to my inexpressible joy, found it spotted, in several places, with what appeared to be figures arranged in lines. Again I placed it in the pan, and suffered it to remain another minute. On taking it off, the whole was just as you see it now."

Here Legrand, having re-heated the parchment, submitted it to my inspection. The following characters were rudely traced, in a red tint, between the death's-head and the goat:

```
53‡‡†305))6*;4826)4‡.)4‡;806*;48†8¶60))8
5;1‡(;:‡*8†83(88)5*†;46(;88*96*?;8)*‡(;4
85);5*†2:*‡(;4956*2(5*_4)8¶8*;4069285);)
6†8)4‡‡;1(‡9;48081;8:8‡1;48†85;4)485†5288
06*81(‡9;48;(88;4(‡?34;48)4‡;161;:188;‡?;
```

"But," said I, returning him the slip, "I am as much in the dark as ever. Were all the jewels of Golconda awaiting me on my solution of this enigma, I am quite sure that I should be unable to earn them."

"And yet," said Legrand, "the solution is by no means so difficult as you might be led to imagine from the first hasty inspection of the characters. These characters, as any one might readily guess, form a cipher—that is to say, they convey a meaning; but then, from what is known of Kidd, I could not suppose him capable of constructing any of the more abstruse cryptographs. I made up my mind, at once, that this was of a simple species—such, however, as would appear, to the crude intellect of the sailor, absolutely insoluble without the key."

"And you really solved it?"

"Readily; I have solved others of an abstruseness ten thousand times greater. Circumstances, and a certain bias of mind, have led me to take interest in such riddles, and it may well be doubted whether human ingenuity can construct an enigma of the kind which human ingenuity may not, by proper application, resolve. In fact, having once established connected and legible characters, I scarcely gave a thought to the mere difficulty of developing their import.

"In the present case—indeed in all cases of secret writing—the first question regards the *language* of the cipher; for the principles of solution, so far, especially, as the more simple ciphers are concerned, depend on, and are varied by, the genius of the particular idiom. In general, there is no alternative but experiment (directed by probabilities) of every tongue known to him who attempts the solution, until the true one be attained. But, with the cipher now before us, all difficulty is removed by the signature. The pun on the word 'Kidd' is appreciable in no other language than the English. But for this consideration I should have begun my attempts with the Spanish and French, as the tongues in which a secret of this kind would most naturally have been written by a pirate of the Spanish main. As it was, I assumed the cryptograph to be English.

"You observe there are no divisions between the words. Had there been divisions, the task would have been comparatively easy. In such case I should have commenced with a collation and analysis of the shorter words, and, had a word of a single letter occurred, as is most likely, (*a* or *I*, for example,) I should have considered the solution as assured. But, there being no division, my first step was to ascertain the predominant letters, as well as the least frequent. Counting all, I constructed a table, thus:

Of the character	8	there are	33.
"	;	"	26.
"	4	"	19.
"	$ and)	"	16.
"	*	"	13.
"	5	"	12.
"	6	"	11.
"	† and 1	"	8.
"	0	"	6.
"	9 and 2	"	5.
"	: and 3	"	4.
"	?	"	3.
"	¶	"	2.
"	] - .	"	1

"Now, in English, the letter which most frequently occurs is *e*. Afterwards, the succession runs thus: *a o i d h n r s t u y c f g l m w b k p q x z*. *E* however predominates so remarkably that an individual sentence of any length is rarely seen, in which it is not the prevailing character.

"Here, then, we have, in the very beginning, the groundwork for something more than a mere guess. The general use which may be made of the table is obvious—but, in this particular cipher, we shall only very partially require its aid. As our predominant character is 8, we will commence by assuming it as the *e* of the natural alphabet. To verify the supposition, let us observe if the 8 be seen often in couples—for *e* is doubled with great frequency in English—in such words, for example, as 'meet,' 'fleet,' 'speed,' 'seen,' 'been,' 'agree,' &c. In the present instance we see it doubled no less than five times, although the cryptograph is brief.

"Let us assume 8, then, as *e*. Now, of all *words* in the language, 'the' is most usual; let us see, therefore, whether there are not repetitions of any three characters, in the same order of collocation, the last of them being 8. If we discover repetitions of such letters, so arranged, they will most probably represent the word 'the.' On inspection we find no less than seven such arrangements, the characters being ;48. We may, therefore, assume that the semicolon represents *t*, that 4 represents *h*, and that 8 represents *e*—the last being now well confirmed. Thus a great step has been taken.

"But, having established a single word, we are enabled to establish a vastly important point; that is to say, several commencements and terminations of other words. Let us refer, for example, to the last instance but one, in which the combination ;48 occurs—not far from the end of the cipher. We know that the semicolon immediately ensuing is the commencement of a word, and, of the six characters succeeding this 'the,' we are cognizant of no less than five. Let us set these characters down, thus, by the letters we know them to represent, leaving a space for the unknown

```
t eeth.
```

"Here we are enabled, at once, to discard the '*th*,' as forming no portion of the word commencing with the first *t*; since, by experiment of the entire alphabet for a letter adapted to the vacancy we perceive that no word can be formed of which this *th* can be a part. We are thus narrowed into

```
t ee,
```

and, going through the alphabet, if necessary, as before, we arrive at the word 'tree,' as the sole possible reading. We thus gain another letter, *r*, represented by (, with the words 'the tree' in juxtaposition.

"Looking beyond these words, for a short distance, we again see the combination of ;48, and employ it by way of *termination* to what immediately precedes. We have thus this arrangement:

```
the tree;4(‡?34 the,
```

or, substituting the natural letters, where known, it reads thus:

```
the tree thr‡?3h the.
```

"Now, if, in place of the unknown characters, we leave blank spaces, or substitute dots, we read thus:

```
the tree thr...h the,
```

when the word '*through*' makes itself evident at once. But this discovery gives us three new letters, *o*, *u* and *g*, represented by ‡, ? and 3.

"Looking now, narrowly, through the cipher for combinations of known characters, we find, not very far from the beginning, this arrangement,

```
83(88, or egree,
```

which, plainly, is the conclusion of the word 'degree,' and gives us another letter, *d*, represented by †.

"Four letters beyond the word 'degree,' we perceive the combination

;46(;88*.

"Translating the known characters and representing the unknown by dots, as before, we read thus:

th.rtee.,

an arrangement immediately suggestive of the word 'thirteen,' and again furnishing us with two new characters, *i* and *n*, represented by 6 and *.

"Referring, now, to the beginning of the cryptograph, we find the combination,

53‡‡†

"Translating, as before, we obtain

.good,

which assures us that the first letter is A, and that the first two words are 'A good.'

"To avoid confusion, it is now time that we arrange our key, as far as discovered, in a tabular form. It will stand thus:

S	represents	a
†	"	d
8	"	e
3	"	g
4	"	h
6	"	i
*	"	n
‡	"	o
(	"	r
;	"	t

"We have, therefore, no less than ten of the most important letters represented, and it will be unnecessary to proceed with the details of the solution. I have said enough to convince you that ciphers of this nature are readily soluble, and to give you some insight into the *rationale* of their development. But be assured that the specimen before us appertains to the very simplest species of cryptograph. It now only remains to give you the full translation of the characters upon the parchment, as unriddled. Here it is:

> *"'A good glass in the bishop's hostel in the devil's seat twenty-one degrees and thirteen minutes northeast and by north main branch seventh limb east side shoot from the left eye of the death's-head a bee line from the tree through the shot fifty feet out."'*

"But," said I, "the enigma seems still in as bad a condition as ever. How is it possible to extort a meaning from all this jargon about 'devil's seats,' 'death's-heads,' and 'bishop's hostels?' "

"I confess," replied Legrand, "that the matter still wears a serious aspect, when regarded with a casual glance. My first endeavor was to divide the sentence into the natural division intended by the cryptographist."

"You mean, to punctuate it?"

"Something of that kind."

"But how was it possible to effect this?"

"I reflected that it had been a point with the writer to run his words together without division, so as to increase the difficulty of solution. Now, a not over-acute man, in pursuing such an object, would be nearly certain to overdo the matter. When, in the course of his composition, he arrived at a break in his subject which would naturally require a pause, or a point, he would be exceedingly apt to run his characters, at this place, more than usually close together. If you will observe the MS., in the present instance, you will easily detect five such cases of unusual crowding. Acting on this hint, I made the division thus:

> *"'A good glass in the Bishop's hostel in the Devil's seat—twenty-one degrees and thirteen minutes—northeast and by north—main branch seventh limb east side—shoot from the left eye of the death's-head—a beeline from the tree through the shot fifty feet out."'*

"Even this division," said I, "leaves me still in the dark."

"It left me also in the dark," replied Legrand, "for a few days; during which I made diligent inquiry, in the neighborhood of Sullivan's Island, for any building which went by the name of the 'Bishop's Hotel;' for, of course, I dropped the obsolete word 'hostel.' Gaining no information on the subject, I was on the point of extending my sphere of search, and proceeding in a more systematic manner, when, one morning, it entered into my head, quite suddenly, that this 'Bishop's Hostel' might have some reference to an old family, of the name of Bessop, which, time out of mind, had held possession of an ancient manor-house, about four miles to the northward of the Island. I accordingly went over to the plantation, and reinstituted my inquiries among the older negroes of the place. At length one of the most aged of the women said that she had heard of such a place as *Bessop's Castle*, and thought that she could guide me to it, but that it was not a castle, nor a tavern, but a high rock.

"I offered to pay her well for her trouble, and, after some demur, she consented to accompany me to the spot. We found it without much difficulty, when, dismissing her, I proceeded to examine the place. The 'castle' consisted of an irregular assemblage of cliffs and rocks—one of the latter being quite remarkable for its height as well as for its insulated and artificial appearance. I clambered to its apex, and then felt much at a loss as to what should be next done.

"While I was busied in reflection, my eyes fell upon a narrow ledge in the eastern face of the rock, perhaps a yard below the summit on which I stood. This ledge projected about eighteen inches, and was not more than a foot wide, while a niche in the cliff just above it, gave it a rude resemblance to one of the hollow-backed chairs used by our ancestors. I made no doubt that here was the 'devil's seat' alluded to in the MS., and now I seemed to grasp the full secret of the riddle.

"The 'good glass,' I knew, could have reference to nothing but a telescope; for the word 'glass' is rarely employed in any other sense by seamen. Now here, I at once saw, was a telescope to be used, and a definite point of view, *admitting no variation*, from which to use it. Nor did I hesitate to believe that the phrases, 'twenty-one degrees and thirteen minutes,' and 'northeast and by north,' were intended as directions for the levelling of the glass. Greatly excited by these discoveries, I hurried home, procured a telescope and returned to the rock.

"I let myself down to the ledge, and found that it was impossible to retain a seat on it unless in one particular position. This fact confirmed my preconceived idea. I proceeded to use the glass. Of course the 'twenty-one degrees and thirteen minutes' could allude to nothing but elevation above the visible horizon, since the horizontal direction was clearly indicated by the words, 'northeast and by north.' This latter direction I at once established by means of a pocket-compass; then, pointing the glass as nearly at an angle of twenty-one degrees of elevation as I could do it by guess, I moved it cautiously up or down, until my attention was arrested by a circular rift or opening in the foliage of a large tree that overtopped its fellows in the distance. In the centre of this rift I perceived a white spot, but could not, at first, distinguish what it was. Adjusting the focus of the telescope, I again looked, and now made it out to be a human skull.

"On this discovery I was so sanguine as to consider the enigma solved for the phrase 'main branch, seventh limb, east side,' could refer only to the position of the skull on the tree, while 'shoot from the left eye of the death's-head' admitted, also, of but one interpretation, in regard to a search for buried treasure. I perceived that the design was to drop a bullet from the left eye of the skull, and that a bee-line or, in other words, a straight line, drawn from the nearest point of the trunk through 'the shot,' (or the spot where the bullet fell,) and thence extended to a distance of fifty feet, would indicate a definite point—and beneath this point I thought it at least possible that a deposit of value lay concealed.

"All this," I said; "is exceedingly clear; and, although ingenious, still simple and explicit. When you left the Bishop's Hotel; what then?"

"Why, having carefully taken the bearings of the tree, I turned homewards. The instant that I left 'the devil's seat,' however; the circular rift vanished; nor could I get a glimpse of it afterwards, turn as I would. What seems to me the chief ingenuity in this whole business, is the fact (for repeated experiment has convinced me it *is* a fact) that the circular opening in question is visible from no other attainable point of view than that afforded by the narrow ledge on the face of the rock.

"In this expedition to the 'Bishop's Hotel' I had been attended by Jupiter, who had, no doubt, observed, for some weeks past, the abstraction of my demeanor, and took especial care not to leave me alone. But, on the next day, getting up very early, I contrived to give him the slip, and went into the hills in search of the tree. After much toil I found it. When I came home at night my valet proposed to give me a flogging. With the rest of the adventure I believe you are as well acquainted as myself."

"I suppose," said I, "you missed the spot, in the first attempt at digging, through Jupiter's stupidity in letting the bug fall through the right instead of through the left eye of the skull."

"Precisely. This mistake made a difference of about two inches and a half in the 'shot'—that is to say, in the position of the peg nearest the tree; and had the treasure been *beneath* the 'shot,' the error would have been of little moment; but 'the shot,' together with the nearest point of the tree, were merely two points for the establishment of a line of direction; of course the error, however trivial in the beginning, increased as we proceeded with the line, and by the time we had gone fifty feet, threw us quite off the scent. But for my deep-seated convictions that treasure was here somewhere actually buried, we might have had all our labor in vain."

"I presume the fancy of *the skull*, of letting fall a bullet through the skull's eye was suggested to Kidd by the piratical flag. No doubt he felt a kind of poetical consistency in recovering his money through this ominous insignium."

"Perhaps so; still I cannot help thinking that common-sense had quite as much to do with the matter as poetical consistency. To be visible from the devil's seat, it was necessary that the object, if small, should be white; and there is nothing like your human skull for retaining and even increasing its whiteness under exposure to all vicissitudes of weather..."

"Yes, I perceive; and now there is only one point which puzzles me. What are we to make of the skeletons found in the hole?"

"That is a question I am no more able to answer than yourself. There seems, however, only one plausible way of accounting for them—and yet it is dreadful to believe in such atrocity as my suggestion would imply. It is clear that Kidd—if Kidd indeed secreted this treasure, which I doubt not—it is clear that he must have had assistance in the labor. But, the worst of this labor concluded, he may have thought it expedient to remove all participants in his secret. Perhaps a couple of blows with a mattock were sufficient, while his coadjutors were busy in the pit; perhaps it required a dozen—who shall tell?"

21.3 Permutation Codes

This is another class of codes that have been in use for a long time. In a *permutation code*, the symbols that comprise the original message are left as they are, but their order is altered according to a prearranged pattern.

Here's an example. Suppose we want to send a message, say "the pen of my aunt is on the table of my uncle." We would start by breaking the message up into groups of 15 characters, not counting spaces (the choice of 15 is arbitrary; its significance will be clear in a moment). In this case, the first group would be

T H E P E N O F M Y A U N T I.

We write out these 15 letters in the squares of a 3×5 rectangular grid, going left to right and top to bottom:

T	H	E	P	E
N	O	F	M	Y
A	U	N	T	I

Now we read off the letters top to bottom and left to right (as for example in a Chinese text), coming up the sequence

T N A H O U E F N P M T E Y I.

which is our encoded message. To decode it, the recipient simply carries out these steps in reverse, entering the letters in order into a 3×5 grid top to bottom and then left to right; then reading the message off left to right and top to bottom.

This was in fact the method of encoding used by the Spartan military, starting in the fifth century B.C. To encode a message, they would take a long strip of parchment and wrap it around a wooden staff. They would then write the message in ordinary fashion—that is, left to right and top to bottom—and then unroll the parchment. Thus, the message

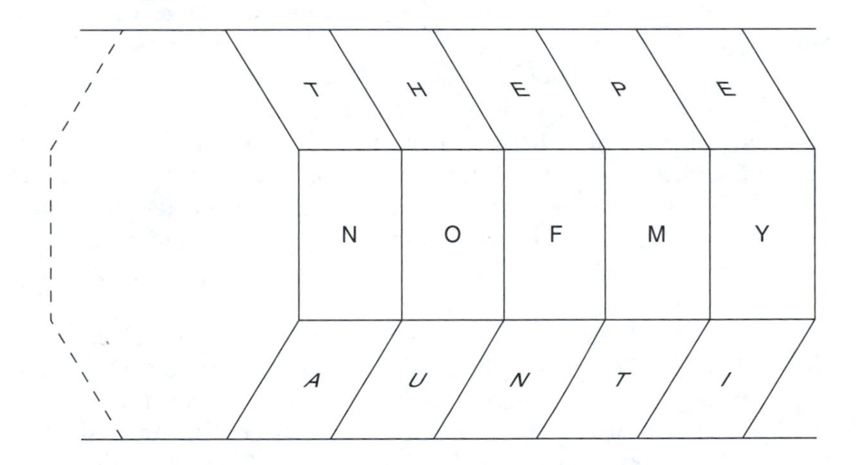

would appear on the strip of parchment as

T	N	A	H	O	U	E	F	N	P	M

The strip would then be carried to the recipient, who would wrap it around his staff to read the message; if the strip were intercepted en route, it would appear to be a meaningless jumble of letters. It doesn't seem like this particular code would keep your average Athenian guessing for more than a few minutes, but apparently this was considered hot stuff back then.

The problem with permutation codes is that, like straight substitution codes, they are relatively easy to crack (especially with a computer), assuming the rearrangement of the letters follows some pattern. For example, if we are trying to decode a message that we suspect has been encoded with a permutation code we look first at the strings of characters formed by taking every other letter; then at the string formed by every third letter, and so on, to see if any of these yield English words or fragments. Even if the rearrangement used to scramble the letters is more complicated, eventually (especially if the message is long enough) we can figure it out by looking for substrings that form words.

21.4 Le Chiffre Indéchiffrable

As you might expect, as people figured out more and more effective ways of deciphering straight substitution codes, codemakers devised ever more clever ways of encoding messages. One key idea was the notion of a *positional substitution code*: basically, a substitution code, but one where the rule we use to substitute for each letter depends on the position of that letter in the message.

For example, suppose at the outset that we choose five different substitution codes; call them Code I, Code II and so on. They need not be related to each other, though in practice—and in the example we'll give in a moment—they will be, just to make life somewhat simpler for the encoder and decoder. We then take the message we want to send and encode the first letter according to Code I, the second letter according to Code II, and so forth. For the sixth letter we go back to Code I, for the seventh, Code II, and so on cyclically through the message.

Of course, there are many variations on this theme: we could use any number of codes rather than five, for example; and we could apply them in any (prearranged) pattern, rather than cyclically. But we'll focus here on one famous example of such a code, the *Vigenère cipher*, named after its inventor, Blaise de Vigenère, a French diplomat of the 15th century. The Vigenère cipher was called "Le chiffre indéchiffrable" when it was introduced; and while this description proved ultimately to be inaccurate it was some centuries before an effective way of cracking it was devised.

Actually, among positional substitution codes, the Vigenère code is pretty simple: all the individual substitution codes used are simply shift substitutions, and they repeat cyclically. Here's how it works. First of all, the key—the information that specifies which codes we are to use—consists simply of a single word or phrase; for an example, let's say the key is the word "BLEAT." To encode a message like "The pen of my aunt is on the table of my uncle," we write it out and under it we write the key word or phrase over and over:

$$\begin{array}{cccccccccccc} T & H & E & P & E & N & O & F & M & Y & A & U\ldots \\ B & L & E & A & T & B & L & E & A & T & B & L\ldots \end{array}$$

The second step of the process is to convert all the letters appearing here into numbers between 1 and 26 according to their position in the alphabet:

$$\begin{array}{cccccccccccc} 20 & 8 & 5 & 16 & 5 & 14 & 15 & 6 & 13 & 25 & 1 & 21 \ldots \\ 2 & 12 & 5 & 1 & 20 & 2 & 12 & 5 & 1 & 20 & 2 & 12 \ldots \end{array}$$

Next comes the essential encoding step: we take each pair of numbers and add them in arithmetic mod 26:

$$\begin{array}{ccccccccccccc} & 20 & 8 & 5 & 16 & 5 & 14 & 15 & 6 & 13 & 25 & 1 & 21 \ldots \\ + & 2 & 12 & 5 & 1 & 20 & 2 & 12 & 5 & 1 & 20 & 2 & 12 \ldots \\ \hline & 22 & 20 & 10 & 17 & 25 & 16 & 1 & 11 & 14 & 19 & 3 & 7 \ldots \end{array}$$

Finally, we convert the resulting string of numbers between 1 and 26 back into letters: 22 becomes V, 20 becomes T, and so on. The final encoded message thus reads

$$V \quad T \quad H \quad Q \quad Y \quad P \quad A \quad K \quad N \quad S \quad C \quad G \ldots.$$

At the other end, the recipient of the message just performs these steps in reverse: assuming she's been given the key word "BLEAT" she converts the string "VTHQY-PAKNSCG" into the string "22 20 10 17 25 16 1 11 14 19 3 7" of numbers between 1 and 26 and does likewise with the word "BLEAT." She then writes out the two strings of numbers—repeating the five number sequence "2 12 5 1 20" coming from "BLEAT" over and over—and subtracts the second from the first in arithmetic mod 26.

$$\begin{array}{ccccccccccccc} & 22 & 20 & 10 & 17 & 25 & 16 & 1 & 11 & 14 & 19 & 3 & 7 \ldots \\ - & 2 & 12 & 5 & 1 & 20 & 2 & 12 & 5 & 1 & 20 & 2 & 12 \ldots \\ \hline & 20 & 8 & 5 & 16 & 5 & 14 & 15 & 6 & 13 & 25 & 1 & 21 \ldots \end{array}$$

Finally she converts the resulting numbers back into letters to arrive at the decoded message. In effect, the Vigenère code amounts to applying a shift-substitution code to each letter of the original message, where the degree of shift is specified by the corresponding letter of the keyword.

The Vigenère code has a number of positive qualities. It's straightforward to encode and to decode messages, assuming you have some basic facility with arithmetic mod 26 (or a table with the same information). A particular virtue is that the fact that once two people have learned the Vigenère code and have agreed to use it, the key—the data needed to encode and decode—is very simple: instead of having to equip both parties with elaborate tables or coding devices, they need only communicate one word like "BLEAT."

Moreover, the Vigenère code, like any positional substitution code, makes it much more difficult to apply all of the techniques we described earlier for cracking substitution codes. Since the letter *e*, for example, may be encoded in five different ways depending on where it in the message it appears, it isn't possible to determine which symbol represents *e* simply by seeing which occurs most often. Likewise, since adjacent letters are encoded in different codes, repeated letters in the text won't appear as repeated letters in the encoded message.

But it can be cracked. Given a long enough message, even encoded with a Vigenère code, patterns will begin to emerge. For example, we know we won't get much information from simply plotting the relative frequency of the letters in the encoded message: the fact that letters are coded differently depending on their position in the text will tend to dampen out the differences in frequency. But here's something we can do: we can look at every other letter and chart their frequencies; then look at every third letter and chart those, and so on. In the case of the Vigenère code with keyword "BLEAT," for example, we'd see relatively little variation in the frequency with which different symbols appeared in the coded message—until we got to looking at every fifth letter (corresponding to the length of the keyword); then all of a sudden we'd see dramatic differences in frequency.

Of course, all of this presupposes that we have enough coded text to run statistical analyses on: if we had only a page or two of text encoded with a Vigenère code, we'd be stuck. But if we had enough text—if we listened in, for example, over time until we had intercepted a number of messages in the same code—we'd be in business. The bottom line is that even a positional substitution code can eventually be cracked. Or at least that was the bottom line until roughly the middle of the 20th century.

21.5 The New World of Coding

Of course, since the introduction of the Vigenère code, there have been many advances in coding, yielding codes that are ever more difficult to break. By the time of the Second World War teams of codebreakers, including some of the best scientific and mathematical minds in England, were working full time to crack the German Enigma code; their eventual success in doing so is viewed as having played a pivotal role in the outcome of the war.

The breaking of the Enigma code is certainly above all a tribute to the brilliance and perseverance of the codebreakers. (They were also aided by Polish refugees, whose experience in the factory that built the enigma machines helped to reverse engineer the way the code was created.) But it's also a reflection of the fundamental flaw in all the codes we've described so far, and in fact of all codes known before the 1960's: that, given a sufficiently long message (or sufficiently many messages) written in a given code, patterns begin to emerge that analysts can use to break the code.

Now, you might say you could overcome this flaw simply by switching codes frequently. But each of the codes we've described requires that the parties in question communicate in advance to agree on a coding system and a key; and if the communication between the parties isn't secure, how are they going to do that? Imagine, for example, a spy in a foreign country who needs to communicate with his home country: if he uses the same code consistently, it'll be cracked; but how can he change it without returning home?

Up until the 1960s, this was the Achilles' heel of all codes: on the one hand, they required prior arrangement between the two parties who wished to communicate securely; and on the other, if you used the same code long enough it could be cracked. It seemed, in other words, that Poe was right when he said, through his character

Legrand, "It may well be doubted whether human ingenuity can construct an enigma of the kind which human ingenuity may not, by proper application, resolve."

But as it turned out, he wasn't. 40 or so years ago, a new idea in cryptography emerged: that of a *public-key cryptosystem*, so called because the key—that is, the information needed to encode a message—is made public. This system, invented in 1978 by Ron Rivest, Adi Shamir, and Leonard Adelman, has two truly remarkable features: it requires no prior arrangements between the two parties who want to communicate securely; and a third party, who's overheard every word spoken and every message exchanged by the two parties since birth, and who has read every document read by them, still can't crack the code.

How is this possible? Well, we'll show you in the following chapter how we can use the special properties of modular arithmetic to construct such a code; in fact it'll use only the ideas and techniques you've already learned in Part III.

Chapter 22

Public-key Cryptography

22.1 The Set-up

Here's the setup: let's suppose there are two parties who wish to communicate. It's traditional in cryptography to call them Alice and Bob, but if you want you can replace these stock characters with You and The Computer at Amazon.com That's Taking Your Order. There's also a third party who is trying to intercept the messages being passed between Alice and Bob (conventionally called Eve, but you could replace her with The Evil Hacker Who Lusts After Your Credit Card Number). The goal is to enable Alice and Bob to communicate without Eve being able to understand. We'll assume the following conditions:

- Alice and Bob have never spoken to each other before in their lives; and
- Eve can hear every message transmitted between Alice and Bob.

To make matters even harder, we'll assume also that Alice and Bob are in possession of no knowledge that is not publicly available. If we assume, for example, that Alice and Bob have read this book, we have to figure that Eve has read it as well.

Under these conditions, can Alice send Bob a message without Eve learning its contents? How is it possible? We'll show you how in this chapter.

22.2 Modular Arithmetic Reappears

To come right to the point, the essential flaw in all the codes described in the last chapter is that they are *symmetric*: that is, the information needed to encode a message is exactly the same as the information needed to decode it. In the straight substitution code used in *The Gold-Bug*, it's the correspondence between the letters of the alphabet and the symbols used in the code to represent them; in the permutation code we described, it's the size of the rectangle used; in the Vigenère cipher it's the keyword. None of these are any good in the present circumstances: if Alice wants to send a message to Bob, she has to know how to encode it; if she knows how to encode it, Eve must know as well; and if Eve knows how to encode the message she knows how to decode it as well.

We have to break this symmetry. In other words, we seek a code where *it requires more information to decode a message than to encode it in the first place.*

How is this possible? The answer lies in the fact that there are mathematical processes that are like lobster traps: they can be done readily enough, but they can't be undone without further information. In fact, you know such a process already: raising a number to a power in arithmetic mod n. Specifically, we've seen in the last part that

- To find the kth power of a number a in arithmetic mod n, we need only know the numbers a, k and n involved; but
- To find the kth *root* of a number in arithmetic mod n, we need in addition to know $\phi(n)$.

"Wait a second," you might say, "I don't get it: doesn't knowing n mean that we know $\phi(n)$ as well? Wasn't there a recipe for finding $\phi(n)$ back in Section 13.2? You even had the gall to call it a 'simple recipe,' as I recall."

You're perfectly right: in theory, knowing n does determine $\phi(n)$. But even a simple recipe can have obscure ingredients, and that's the case here: *in order to calculate $\phi(n)$, you have to know the prime factorization of n.* (For example, if $n = pq$ is the product of two prime numbers p and q, then as we saw $\phi(n) = (p-1)(q-1)$.) And if n is large enough—several hundred digits long, say—then, as we saw in Section 12.7, factoring n is a calculation that would take longer than the lifetime of the universe. The key fact here, in other words, is this: if you take two very large prime numbers p and q and multiply them to get a number n, you can tell the whole world what n is, but *only you will know p and q*, and so *only you will be able to calculate $\phi(n) = (p-1)(q-1)$.*

"All right, then," you say, "I'm willing to buy that there really is a difference between the information you need to take powers in arithmetic mod n and what you need to take roots mod n. But I still don't see how you can use that fact to design the sort of code you're talking about."

OK, here goes:

22.3 The Return of Bob and Alice (and Eve)

Let's say Alice wants to send Bob a message, under the conditions set out at the beginning of this chapter. We're going to assume that the message is in the form of a number a of a hundred digits or less (if Alice wants to send Bob a text message, she just converts it to numbers using a standard translation, like the ASCII; if the message is longer than 100 digits, she simply breaks it up into a series of 100-digit numbers). Here's how they do it:

First step (Public). Alice sends Bob a plain (uncoded) message, saying "I need to send you some information." Bob's ears perk up; so do Eve's, since she's overhearing all this.

Second step (Private). Bob goes off and secretly finds two prime numbers p and q, each (say) a hundred or two hundred digits long. He then multiplies them together to form a number n, and while he's at it also calculates $\phi(n) =$

$(p-1)(q-1)$. Finally, he chooses any number k that is relatively prime to $\phi(n)$.

Third step (Public). Bob tells Alice (and presumably Eve as well, and anyone else who's interested) the numbers n and k. This is the data Alice needs to encode her message—the "public key" that gives the process its name.

Fourth step (Private). Now comes the crucial step: Alice takes the number a that comprises her message, and raises it to the kth power in arithmetic mod n, to arrive at a new number b. This is the encoded message.

Fifth step (Public). Alice sends the encoded message, consisting of the number b, to Bob (and Eve).

Sixth step (Private). Bob takes the number b that Alice has transmitted, and, *using his knowledge of* $\phi(n)$, finds the kth root of b in arithmetic mod n. The resulting number a is the original message that Alice wanted to transmit.

And that's it. As improbable as it may have sounded a page or two ago, we've now done exactly what we promised. Alice has sent the message to Bob and, assuming both are up on their modular arithmetic, it's been received and decoded. Moreover Eve, who's heard every exchange (and who's read this book) is gnashing her teeth in frustration: she knows the modulus n, she knows the power k, she knows the encoded message b and knows that b is just the kth power of the actual message a in arithmetic mod n—but she can't decode the message. Without knowing how to factor n as a product of primes, she doesn't know $\phi(n)$; and without knowing $\phi(n)$ she doesn't know how to find the kth root of b in arithmetic mod n.

In other words: everybody—Alice, Bob, Eve and your grandmother if she's listening in—knows how to encode the message; *but only Bob knows how to decode it*. This is what we mean by an asymmetrical code. Note that if Alice lost her original message a after encoding it, she wouldn't be able to decode it herself! To top it all off, if Alice and Bob want to continue the conversation—if, say, Bob wants to reply to Alice—they can repeat this process in full. In other words, *they can use a different key for each communication back and forth.*

Here's a diagram of the process by which Alice transmits the message a to Bob:

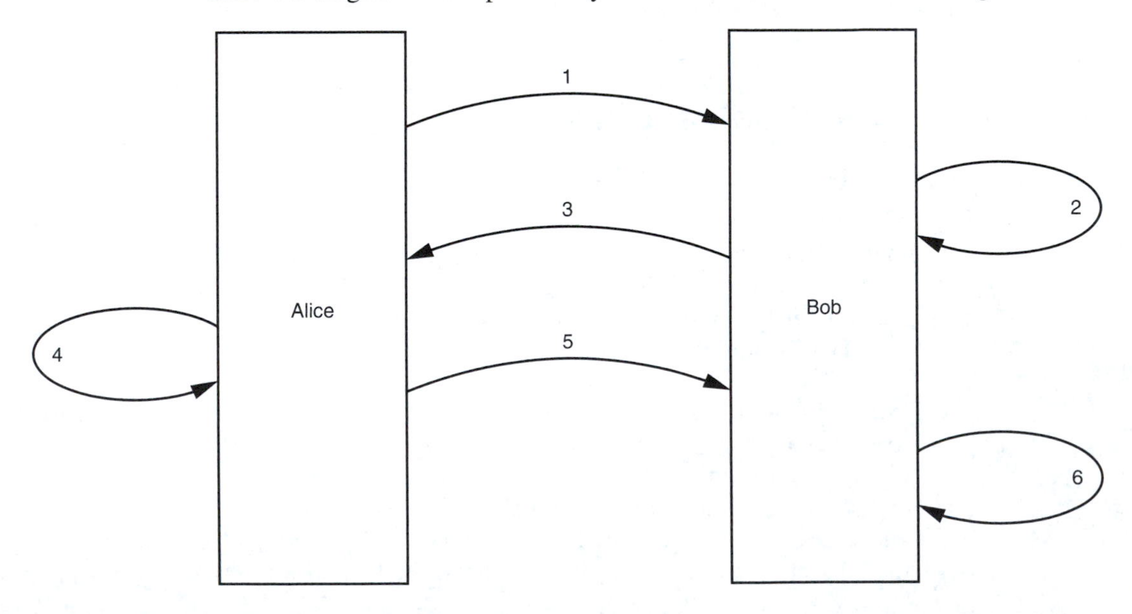

1. Alice to Bob: "Heads up!"
2. Bob: chooses p, q; calculates $n = pq$; chooses k
3. Bob to Alice: "n, k"
4. Alice: calculates $b \equiv a^k \pmod{n}$
5. Alice to Bob: "b"
6. Bob: calculates $a \equiv \sqrt[k]{b} \pmod{n}$

Now you try it. Of course, the numbers n, k, a and b used in an actual secure message would be too large to work with readily (after all, the security of the code depends on Eve's inability to factor n), so this exercise uses smaller numbers.

Exercise 22.3.1 You are an FBI agent who has just intercepted a message from mobster Tony Soprano to his enforcer, Peter "Paulie Walnuts" Gualtieri. Tony and Paulie are using the public-key modular arithmetic encryption system with modulus $n = 437$ and exponent $k = 61$, so that to encode his message Tony raises it to the 61st power (mod 437). The encoded message that you intercept is "9".

Tony and Paulie, however, have sadly underestimated today's FBI! Working overtime, a team of crack FBI cryptographers has determined that $437 = 19 \times 23$. Now crack Tony's message (that is, find the 61st root of 9 (mod 437)), showing all your steps.

22.4 FAQ

Q. OK, this all sounds pretty good if you've got the math. But what do the other 99% of us do?

A. Actually, we all use this code, or variants of it, pretty much every day, even when we're not aware of it. When we go the ATM, when we log in to get e-mail from a remote location, when we order stuff online—these codes, or variants of them, are used to transmit information securely.

Q. What are you talking about? I've never heard of this code before today, so how could I have been using it?

A. It's built in to all the standard Web browsers and e-mail programs.

Q. Well then, if my browser carries all this out automatically and I'm willing to trust it, why do I have to learn how it works?

A. Because you have a healthy curiosity about the world around you and the role that mathematics plays in it. That's why you're reading this book, remember?

Q. It seems like the security of this code depends completely on the assumption that Eve can't factor n. What happens if she figures out a way to factor 300-digit numbers?

A. In that case, we—and civilization as we know it—are in deep, deep trouble. Basically, all secure transmissions—financial transfers between banks, military and industrial secrets—use variants of this procedure for privacy and verification. Anyone in possession of an algorithm for factoring large numbers in reasonably short time would have effectively universal access.

Q. But that couldn't happen, right? I mean, mathematicians have proved that there can't be a procedure for factoring large numbers in reasonably short time, haven't they?

A. Well, ummm . . . not exactly. No one has ever found such an algorithm, and lots of people have thought about the problem. But, no, it hasn't actually been proved that one doesn't exist.

This is the sort of problem that computing has brought into mathematics, and it has a different flavor. Here we are trying to prove that something which is theoretically simple to understand is difficult to accomplish in practice. At present, it's one of the major unsolved problems in mathematics.

Q. So what you're saying is, some nerd figures out how to factor large numbers and the next time I go to the bank I find my life's savings have been transferred to Uzbekistan?

A. Ummmm . . . Don't you have any more questions about the mathematics?

Q. Oh all right. Here's one: In the second step of your description of the code, you say "Bob goes off and secretly finds two prime numbers p and q, each (say) a hundred or two hundred digits long." How do we know there even are primes as big as that?

A. That's better. The answer is, we proved that there are infinitely many primes. So no matter how big a number you pick—10^{100} or whatever—there're guaranteed to be primes larger than that. Infinitely many of them, in fact.

Q. OK, I believe there are primes as big as you like. But how does Bob find one? I wouldn't have any idea where to begin finding a large prime number like that. Come to think of it, I wouldn't even know how to tell if a number that big was prime or not, unless it was even or ended in a 5 or something.

A. Now, that's a great question. In fact, that's what we're going to be discussing in the next chapter, so stick around and you'll see the answer.

HA HA. I WROTE A SECRET MESSAGE AND YOU CAN'T READ IT!
WHAT'S IT SAY?
©1988 Universal Press Syndicate
YOU'LL NEVER KNOW.
GIMME THAT.
THE CAPTAIN MARVELOUS CODE EMPLOYS THE MOST SOPHISTICATED ENCRYPTION ALGORITHMS EVER DEVELOPED. IT CAN'T BE CRACKED.
IT SAYS "PAIGE IS A FATHEAD."
AMEND
THAT WAS A LUCKY GUESS, RIGHT? I MEAN IT HAD TO BE A LUCKY GUESS...
YOU'LL NEVER KNOW.
11-29

Chapter 23

Finding Primes

In this chapter, we're going to return to the discussion we began in Part II about the distribution of primes. The mystery of how and where primes occur among all numbers is one of awesome subtlety and beauty, full of questions that are as easy to ask and simultaneously as difficult to resolve as anything in mathematics. But we're not going to address these issue directly here. Rather, our focus in this chapter will be on a fairly simple and down-to-earth question: *how do we actually find prime numbers?*

We've already shown in Section 10.5 that there are infinitely many prime numbers. So if, for example, you were asked to come up with a prime number at least 100 digits long, you could be confident that an answer existed: if all primes were less than 10^{100}, there could be only finitely many of them. But knowing that a prime with more than 100 digits exists is one thing; finding one is another. In fact, the method commonly used to find large primes is based on modular arithmetic, as we'll see by the end of this chapter.

23.1 How Frequently Do Primes Occur?

If we're going to go hunting primes, we should know something about their habitat: how common or sparse they are, where they like to hang out, and so on. Again, the theorem we proved in Section 10.5 says that there exist primes larger than 10^{100}—infinitely many of them, in fact—but doesn't guarantee that there are any in any specified range. For all we know, in other words, there could be no primes at all between 10^{100} and 10^{200}. To embark on an expedition to find primes in this range without first asking whether we might reasonably expect to find any could turn out to be like organizing an expedition to locate penguins in the Sahara.

Our first mission, then, is to give some sort of answer to the question, "what fraction of all numbers are prime?" Actually, that's our second mission; our first mission would be to make sense of this question, since taken literally it's meaningless. After all, there are infinitely many numbers, and infinitely many primes, so it doesn't make sense to talk about "the fraction of all numbers that are prime."

To make the question more meaningful, here's what we can do: first of all, we can look at the numbers from 1 to, say, 10, and ask, "what fraction of these numbers are prime?" We'll call the answer to this x_1. Next, we look at all numbers from 1 to 100, and ask the same question; we call this fraction x_2. Then we do it for the numbers between 1 and 1,000, between 1 and 10,000, and so on. We ask: how do these numbers $x_1, x_2, x_3, \ldots$ behave as the numbers involved get large?

As you might expect, people have studied this question in all sorts of ways. On a theoretical plane, some of the most beautiful (and the most complex) mathematics of the 19th and 20th centuries was concerned with exactly this matter. Both the methods used and the statements of the results obtained are beyond our present scope, but we can certainly give you some of the empirical data that people have found.

In fact, people have found all the primes up to 10,000,000,000 (and beyond, but that's as far as we'll go here); and we can correspondingly answer the question "what fraction of all numbers between 1 and N are prime?" for all numbers N up to 10,000,000,000. We give some of the results here in the table below. The second column of this table gives the number of prime numbers less than 10^n (that is, with n or fewer digits) for $n = 1, 2, \ldots, 10$, and the third column gives the proportion of all numbers with n or fewer digits that are prime.

number N	*number of primes less than N*	*probability that a number picked at random between 1 and N will be prime*
10	4	1 in 2.5
100	25	1 in 4
1,000	168	1 in 5.952...
10,000	1,229	1 in 8.137...
100,000	9,592	1 in 10.425...
1,000,000	78,498	1 in 12.739...
10,000,000	664,579	1 in 15.047...
100,000,000	5,761,455	1 in 17.357...
1,000,000,000	50,847,534	1 in 19.667...
10,000,000,000	455,052,512	1 in 21.976...

This table is fascinating in a number of ways. The first thing we might observe is the simple fact that primes get more and more sparse as numbers get larger and larger: roughly 1 in 6 numbers less than 1,000 are prime, while less than 1 in 10 numbers with at most five digits are prime, and less than 1 in 15 numbers with seven or fewer digits are. This is not so surprising, after all, if we think of the sieve of Eratosthenes: to find primes less than 10^4, for example, we have to check divisibility by all primes up to 100, whereas to find primes less than 10^8 we have to check divisibility by all primes up to 10,000. In other words, we have to apply the sieve more times for the larger numbers, which makes it plausible at least that the proportion of primes among all numbers less than 100,000,000 should be smaller than the proportion of primes among all numbers less than 10,000.

On the other hand, we see that while the fraction of all numbers less than N that are prime gets smaller as N gets larger, it does so very gradually. Each time we multiply the number N by 10, the odds against a number picked at random among numbers less than N lengthen, but not dramatically: the numbers in the third column of our table are growing at a measured pace, increasing less than 3 each time.

In fact, when we look at the numbers in the third column of the table and how they increase we begin to see a truly remarkable pattern. To bring this out, let's make a table of the numbers in the third column and their successive differences, that is, how much larger each is than the one before:

2.5	
4	1.5
5.952...	1.952...
8.137...	2.185...
10.425...	2.288...
12.739...	2.314...
15.047...	2.308...
17.357...	2.310...
19.667...	2.310...
21.976...	2.309...

What we see from this is that the differences seem to be remarkably consistent, after the first few. This has fascinated mathematicians for centuries, and as a result of their study we have learned that in fact this pattern persists forever, with the differences approaching a constant equal to

$$\Upsilon = 2.30258509299\ldots$$

This fact is a famous result in number theory, called the *Prime Number Theorem*:

> The proportion of all numbers with n or fewer digits that are prime is approximately $1 : \Upsilon n$, where Υ is the constant above; and the approximation gets better as n grows larger.

23.2 Many, Many Questions

It's one of the most annoying things about mathematics: every time you think you've actually solved a problem and proved a theorem, it turns out you've just raised a half a dozen new questions. This is no exception: seeing the above statement, we immediately want to know a number of things:

- Does the number Υ have a precise definition, or just a decimal approximation?
- How good is the approximation $1 : \Upsilon n$ to the actual proportion of numbers with n or fewer digits that are prime?
- Is there a more accurate approximation?

Whole realms of mathematics have grown up out of our attempts to solve these and similar problems, but to say more would go way beyond the scope of this book. (One thing we should say is that the number Υ does have a precise definition; it's the logarithm of 10 to the base e, where e is the constant appearing in calculus.) In fact, there are more accurate approximations, but for our purpose, which is simply to get a rough sense of how many primes there among numbers of a certain size, the approximation here is more than good enough. For example, the Prime Number Theorem asserts that the odds that a 200-digit number picked at random is prime are approximately one in $200\Upsilon \sim 460$; the actual odds will fall somewhere between one in 450 and one in 470.

23.3 How Primes Are Found

At this point, we are ready to describe the procedure by which large primes are found. To heighten the drama, pause for a moment and think about what sort of mathematical processes might be employed; and below we'll reveal to you the method, based on centuries of study of number theory, that is used millions of times every day to find large prime numbers:

We guess.

The fact is, there is no known way to generate arbitrarily large primes; the best that people can do is simply to pick a number at random and hope. It's not as loopy as it sounds: based on what we said in the last section, if we simply write out a string of, say 100 random digits, the odds of hitting a prime are roughly one in 230 (and we can improve those to something like one in 60 by simply avoiding even numbers and numbers divisible by 3 or 5). If we choose a hundred or so such numbers, the odds are in our favor that at least one will be prime.

But now we come up against a basic fact of life, one that's important enough to merit its own box:

If you're going to guess the answer to a problem, you need a way of telling when you're right.

In other words, how do we go about determining whether the 100-digit number n we've just pulled out of our hat is prime or not?

Now, if we asked you to say whether a number like, say, 323 was prime, the procedure would be clear: check if it's divisible by 2 (it's not); check whether it's divisible by 3 (ditto); then check divisibility by 5, 7 and so on. We would just have to check divisibility by prime numbers less than $\sqrt{323}$, that is, up to 17. Even for those of us whose calculator fingers are not especially nimble, it wouldn't take more than a minute or so to do this.

But how long would it take to check a 100-digit number n in this way? We'd have to check divisibility by numbers up to $\sqrt{n}$, which in this case is roughly 10^{50}. And while it's true we only have to check divisibility by primes, this doesn't gladden our hearts all that much, for two reasons. One, we don't have a list of all the primes up to 10^{50}. Two, even if we did, since roughly one out of a 120 numbers in that range are prime, the number of primes less than 10^{50} would be something like 10^{48}. And, as we worked out in Section 12.7, the time it would take us to carry out that many operations is several orders of magnitude longer than the lifetime of the universe.

Is there some way to tell whether our chosen 100-digit number is prime or not before your grandchildren's grandchildren are eligible for social security? In fact there is, based on Fermat's Theorem; and in the remainder of this chapter we'll show you how it works. This—or rather a refinement of it, which we'll also describe—is actually how people find large prime numbers: they guess, and test, until they get lucky.

23.4 Mathematicians and Logic

Before we tell you how to test whether a large number is prime or not, we should say a few words about mathematicians and logic. Consider the two statements:

"All iguanas are green."

and

"Anything that is not green is not an iguana."

These two statements are exactly equivalent: one can't be true without the other being true, and vice versa. Logically, if we wanted to demonstrate the truth of the first statement, we could do so by proving the second, and vice versa.

In the real world, of course, this is nonsense. Suppose, for example, you wanted to test the first statement above. You might wander around looking for iguanas and checking the color of any you see; if you encounter a hundred iguanas and discover that they're all green, you could reasonably take that as evidence of the truth of the first assertion. You wouldn't have proved it, certainly—for all you know, there could be a blue iguana right around the corner—but anyone would consider it at least plausible, on the basis of your evidence, that the statement was true.

But suppose instead you were to apply the same test to the second statement. That is, you go around looking for nongreen objects, and checking to see if any of them are iguanas. You make a list of the first hundred nongreen objects you see—your shirt, your socks, your pen, etc.—and determine that none of them are iguanas. Nobody in his right mind would say you had shown that all iguanas are green, or even found meaningful evidence. In other words, a reasonable person would not equate the two statements above.

But mathematicians are not reasonable people! Manipulating the truth is what we do for a living, and we couldn't think of one of the statements above without thinking of the other.

With that in mind, let's go back and look again at Fermat's Theorem. It says that

If n is any prime number and a any number between 1 and $n-1$, then

$$a^{n-1} \equiv 1 \pmod{n}.$$

Now, that seems like a completely clear statement (once you get past the fact you need to learn modular arithmetic to make sense of it). But to a mathematician it means many different things: for example, it is logically equivalent to the formulation

If n is any number and a any number between 1 and $n-1$, and

$$a^{n-1} \not\equiv 1 \pmod{n},$$

then n is not prime.

This, if you think about it, is exactly the answer to our question.

23.5 Is 21 a Prime?

Let's suppose we're set the task of determining whether or not the number 21 is prime. We leap into action: "Let's see," we say, "Fermat tells us that, if 21 *were* prime, then we could take any number between 1 and 20, raise it to the 20th power in arithmetic mod 21, and we'd get the answer 1. Let's try it with the number 2." We then calculate 2^{20} (mod 21), in the usual way: we make a table of powers of 2 (mod 21) by successive squaring

$$\begin{aligned} 2^2 &\equiv 4 \\ 2^4 &\equiv 4^2 \equiv 16 \\ 2^8 &\equiv 16^2 \equiv 4 \\ 2^{16} &\equiv 4^2 \equiv 16 \end{aligned}$$

and then, since $20 = 16 + 4$, we have

$$2^{20} \equiv 2^{16} \cdot 2^4 \equiv 16 \cdot 16 \equiv 4 \pmod{21}$$

"So," we say, "since 2^{20} in arithmetic mod 21 is 4, and not 1, we have now conclusively established that 21 cannot be prime!"

Now this, you might say, is just the sort of jaw-droppingly stupid idea you'd expect from a bunch of mathematicians: any fool can see that 21 isn't prime. Why would you waste your time doing an elaborate calculation in arithmetic mod 21 to tell that?

Of course it's true that it takes almost no time to see that 21 is divisible by 3, whereas it takes a good bit longer to calculate 2^{20} (mod 21) and see that it isn't 1. But suppose on the other hand that we want to test whether the number

$$\begin{aligned} n = {} & 821896234616933360506404669200020103992240 59 \\ & 41911209155466063911044893991089188784552603 \\ & 96293373195506744671107906900359210216326161 \\ & 77020584193395847238632172619539403790193935 \\ & 5001963066620879377553140012133 61 \end{aligned}$$

is prime or not. Strange as it may sound at first, the situation is exactly reversed: as we've seen, even with the fastest computers around today it would take centuries to test whether this number is prime by dividing it by all smaller numbers. By contrast, while it may seem daunting to pick a random number and raise it to the $(n-1)^{\text{st}}$ power (mod n), in fact it involves only (only!) a few hundred calculations, each involving numbers of a few hundred digits—something your laptop can probably do faster than we can type the period at the end of this sentence.

23.6 Can I Get a Witness

Let's formalize the process we've proposed in the last section for prime testing. Suppose we're given a large number n and want to know if it's prime. Here's how we decide: first we pick a number a at random between 1 and $n-1$. The number a will be called a *witness*, for reasons that will be clear in a moment. Now we raise the number a to the $(n-1)^{st}$ power in arithmetic mod n, and see what we get. There are two possibilities:

- If the result a^{n-1} (mod n) is any number other than 1, we may conclude immediately and without doubt that n *cannot be a prime number*. No further tests are necessary.
- If the result a^{n-1} (mod n) *is* equal to 1, then we can't logically conclude anything for certain: n might be a prime; or it might be just a coincidence. In this case, to try to reach a definitive answer we have to test again.

The process thus consists of a repeated testing of the number n: if at any point n fails the test, we can conclude that n can't be prime, and stop; as long as n keeps passing the tests, we can't say whether or not n is prime with absolute certainty—though each time n passes a test, it's less and less likely that it's just a coincidence and increasingly likely that n really is a prime.

Think of it this way: imagine we're in an *Alice in Wonderland* sort of courtroom, with the Queen of Hearts as the judge and the number n as the defendant, accused of the crime of being a product of two smaller numbers. What we'll do is call a witness: that is, a randomly chosen number a between 1 and $n-1$. If a testifies that n isn't prime—in other words, if $a^{n-1} \not\equiv 1$ mod n—that's it; n is obviously guilty and it's "off with his head." And if a testifies that n might very well be innocent—that is, if $a^{n-1} \equiv 1$ mod n, suggesting that n probably is prime—well then, we'll just call another witness.

For the flowchart-minded, here's one:

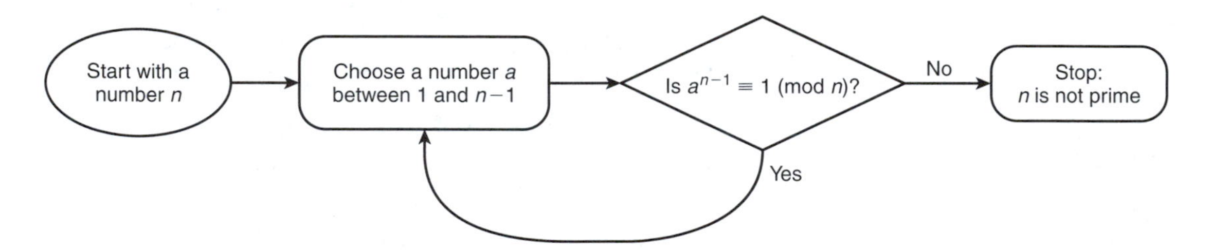

Note that this process doesn't necessarily stop: as long as our witnesses testify that n could be prime, in theory, we keep going. But in practice, if we find that $a^{n-1} \equiv 1$ (mod n) for a large number of witnesses—say, a few hundred values of a—we'd stop and say, "the number n is probably prime." (We'll see in the next section how to quantify that "probably.")

In any event, this is the answer to the question, "How do we find large primes?" Simply put, we pick a number at random and test it for primality; if it fails we try another, while if it passes a number of tests it seems reasonable to assume that it's prime.

23.7 It's Always Something

Now, we can—and probably should—stop here. But there's something worrying about the process we've just described. Specifically: is it possible that a number n could pass the test we've devised with respect to every number a relatively prime to

n but still not be prime? In other words, could there be in our courtroom defendants who are guilty, but who get away with it because there are no witnesses? And if so, are there more accurate tests for primality we could devise?

The answer to both these questions is "yes." To begin with, there are numbers that behave in exactly this fashion—in other words, numbers n that are not prime, but have the property that $a^{n-1} \equiv 1 \pmod{n}$ for all numbers a relatively prime to n. Such numbers are called *Carmichael numbers*.

Here's how they do it: we know that for any number n, and any number a relatively prime to n,

$$a^{\phi(n)} \equiv 1 \pmod{n}.$$

On the other hand, what we're asking when we perform our test is whether it's the case that

$$a^{n-1} \equiv 1 \pmod{n}.$$

Now, if n is not prime, then $\phi(n)$ is certainly not equal to $n-1$, so the first statement wouldn't seem to have much to do with the second; and in general it doesn't. But there is one circumstance in which it would: *if $\phi(n)$ actually divided $n - 1$*. In this case, the first statement would actually imply the second: if $n - 1$ were a multiple of $\phi(n)$—that is, if we could write

$$n - 1 = m \cdot \phi(n)$$

for some whole number m—then it'd also be true that

$$\begin{aligned} a^{n-1} &\equiv (a^{\phi(n)})^m \\ &\equiv 1^m \\ &\equiv 1 \pmod{n}, \end{aligned}$$

so n would pass our test with respect to every a relatively prime to n.

In fact, if we look a little closer, we see that in order for a number n to be a Carmichael number, $\phi(n)$ doesn't actually have to divide $n - 1$. In fact, n will be a Carmichael number if two things are true about n:

- n is a product of distinct primes; that is, in the prime factorization of n, no prime appears to a power greater than 1; and
- for each prime number p dividing n, the number $p - 1$ divides $n - 1$.

Why is this? Well, let's suppose n is a number satisfying these two conditions, and say a is any number relatively prime to n. For each prime p that divides n, we can write

$$n - 1 = m \cdot (p - 1)$$

for some whole number m; and so we see that

$$\begin{aligned} a^{n-1} &\equiv (a^{p-1})^m \\ &\equiv 1^m \\ &\equiv 1 \pmod{p} \end{aligned}$$

Note that we're working mod p here, not mod n. What this says is that

$$a^{n-1} - 1 \equiv 0 \pmod{p},$$

that is, $a^{n-1} - 1$ *is divisible by* p. But if $a^{n-1} - 1$ is divisible by every prime dividing n, and n satisfies the first condition above, then $a^{n-1} - 1$ must be divisible by n itself. In other words, n will seem to be prime according to our test, no matter how many witnesses we call.

Do such numbers actually exist? It's simple enough to ask that "for each prime number p dividing n, the number $p - 1$ divides $n - 1$;" but it's a lot trickier to arrange for it to happen: the numbers that divide $n - 1$ in general have little to do with the numbers dividing n. But Carmichael numbers *do* exist. For example, we could take

$$n = 561.$$

If we factor n, we find that

$$n = 3 \cdot 11 \cdot 17;$$

on the other hand, we see that

$$n - 1 = 560 = 2^4 \cdot 5 \cdot 7.$$

Now, for each of the three primes p dividing n—that is, 3, 11 and 17—we check to see if $p - 1$ divides $n - 1$, and we see they all do: 2, 10 and 16 all divide 560. So 561 is a Carmichael number. (In fact, it's the smallest one.)

Exercise 23.7.1 Check that the numbers 1,105 and 1,729 are also Carmichael numbers. Can you find any others?

Exercise 23.7.2 Can a product of two prime numbers be a Carmichael number?

The upshot in any case is that there are numbers n that will fool our prime test: all witnesses a (except for those prime to n relatively few not) will attest to their possible primality. So we need to refine our test. We need, in other words, to find another property of arithmetic mod n that holds when n is a prime, but not when n is composite.

23.8 How Many Square Roots of 1 Are There in Modular Arithmetic?

The property we'll use is in fact a simple one. To start with, let's ask a very naive question: How many square roots of 1 are there in arithmetic mod n? There's 1 and -1, of course, but are there others?

Let's start by looking at the case of a prime modulus. The first thing to do is to go back and look at the power tables for arithmetic mod 7 and 11 that we worked out in Part III, and the power tables for arithmetic mod 5 and 13 that you worked out yourself back then (you did keep a copy, right?). Scan the "x^2" column: do you see any 1s, except in the first and last row?

We didn't think so. The fact is, *in arithmetic mod a prime number, there are no square roots of* 1 *except* 1 *and* -1. Here's why: suppose p is any prime number, and a is any square root of 1 in arithmetic mod p—that is, a number such that

$$a^2 \equiv 1 \pmod{p},$$

or, in other words,

$$a^2 - 1 \equiv 0 \pmod{p}.$$

This says that p divides $a^2 - 1$. But we can also write $a^2 - 1$ as

$$a^2 - 1 = (a - 1)(a + 1),$$

and it's the fundamental property of prime numbers p that *if p divides a product, it must divide one of the factors*. In other words, if p divides the product $(a-1)(a+1)$, then either p divides $a - 1$ or else p divides $a + 1$. In the first case, $a \equiv 1 \pmod{p}$; and in the second case $a \equiv -1 \pmod{p}$. Thus we see the only possible square roots of 1 in arithmetic mod p are ± 1.

By contrast, things are very different in arithmetic mod n when n is not a prime number. Here a little experimentation yields lots of cases where there are more than two square roots of 1: for example, in arithmetic mod 8 we see that

$$1^2 \equiv 3^2 \equiv 5^2 \equiv 7^2 \equiv 1 \pmod{8};$$

in arithmetic mod 15 we have

$$1^2 \equiv 4^2 \equiv 11^2 \equiv 14^2 \equiv 1 \pmod{15};$$

and in arithmetic mod 21 we have

$$1^2 \equiv 8^2 \equiv 13^2 \equiv 20^2 \equiv 1 \pmod{21}.$$

These examples should convince you that there may be many different square roots of 1 in arithmetic mod n. If you're curious to see more, the following exercise is a little more difficult than most, but should be rewarding.

Exercise 23.8.1 In arithmetic mod 105, there are actually 8 square roots of 1! Can you find them?

(Hint: $105 = 3 \cdot 5 \cdot 7$, so to say that 105 divides $a^2 - 1$ means exactly that 3, 5 and 7 do; in other words, we're looking for numbers a that are congruent to ± 1 modulo each of the primes 3, 5 and 7.)

Once you've done this exercise, you'll be able to see that in general, if a number n is a product of two odd primes then arithmetic mod n will have four square roots of 1; if it's a product of three odd primes then arithmetic mod n will have eight square roots of 1, and so on.

In any event, the bottom line is that we may be able to distinguish between prime numbers and nonprime numbers n by examining the square roots of 1 in arithmetic mod n. But how do we go about looking for such square roots? How do we make an effective test of primality out of this? We'll see how in the following section.

23.9 The Miller-Rabin Test

The Miller-Rabin test is actually a refinement of the Fermat test we've discussed. In effect, it's a way of possibly extracting a little extra information from the witnesses.

Here's how it works, in an example. Suppose we want to test the number $n = 561$ for primality. (Of course we've just seen that 561 is composite, but we want to keep the numbers in this example reasonably small.) In the Fermat test, we'd pick a random number—say, in this case, 7—and raise it to the 560th power in arithmetic mod 561. If the result is not equal to 1, of course, we know immediately that 561 can't be prime. But what if the result is 1, as it is in this case?

Here is where the Miller-Rabin test diverges from the Fermat test. In carrying out the Fermat test, we'd calculate 7^{560} in arithmetic mod 561 the usual way: express 560 as a sum of powers of 2, make a list of powers of 7 by successive squaring, and so on. But here we do it slightly differently. Specifically, they say: first divide 560 by 2 until you get an odd number. In this case, 2 divides 560 a total of 4 times: we can write

$$560 = 35 \times 2^4.$$

Now, instead of finding $7^{560} \pmod{561}$ the usual way, first find $7^{35} \pmod{561}$. Then square it to find $7^{70} \pmod{561}$; square it again to arrive at $7^{140} \pmod{561}$, and continue in this way two more times until we get to $7^{560} \pmod{561}$. Here's a table of the results:

$$\begin{aligned}
7^{35} &\equiv 241 \pmod{561};\\
7^{70} &\equiv 241^2 \equiv 298 \pmod{561};\\
7^{140} &\equiv 298^2 \equiv 166 \pmod{561};\\
7^{280} &\equiv 166^2 \equiv 67 \pmod{561}; \text{ and}\\
7^{560} &\equiv 67^2 \equiv 1 \pmod{561}.
\end{aligned}$$

Now, if our only concern were to calculate $7^{560} \pmod{561}$, this wouldn't be the most efficient way to go about it. But it has one signal virtue. Here we see that $7^{560} \equiv 1 \pmod{561}$, which means that as far as the Fermat test is concerned 561 might well be prime. (Indeed, we knew this would happen: 561 is a Carmichael number, as we saw in Section 23.7.) But, precisely because $7^{560} \equiv 1 \pmod{561}$, *the number 67 immediately preceding* 7^{560} *in this list is a square root of 1.* Since it's not either 1 or -1, we see that 561 can't be prime.

So this is how the Miller-Rabin test goes:

- First, we take $n - 1$ and divide by 2 until we arrive at an odd number: that is, we write $n - 1$ as

  $$n - 1 = 2^k \cdot m$$

  for some whole number k and odd number m.

- Now, we proceed as in the Fermat test. To start with, we pick a witness—that is, a number a between 1 and $n - 1$. We then raise it to the $(n-1)^{\text{st}}$ power in arithmetic mod n, but not in the usual way: instead, we first calculate a^m, and then square it k times to arrive at $a^{n-1} \pmod{n}$. We make a table of the results:

  $$\begin{aligned}
  a^{2m} &\equiv (a^m)^2 \pmod{n};\\
  a^{4m} &\equiv (a^{2m})^2 \pmod{n};\\
  a^{8m} &\equiv (a^{4m})^2 \pmod{n};
  \end{aligned}$$

and so on until we arrive at

$$a^{2^{k-1}m} \equiv (a^{2^{k-2}m})^2 \pmod{n}; \text{ and}$$
$$a^{n-1} = a^{2^k m} \equiv (a^{2^{k-1}m})^2 \pmod{n}.$$

- Now, we examine the sequence of numbers

$$a^m, \quad a^{2m}, \quad a^{4m}, \ldots\ldots, a^{2^{k-2}m}, \quad a^{2^{k-1}m}, \quad a^{2^k m} \equiv a^{n-1}$$

that we've produced in this way. If the last one on the list—that is, $a^{n-1} \pmod{n}$—is anything other than 1, we know by the Fermat test that n can't be prime.

- If, on the other hand, we find that $a^{n-1} \equiv 1 \pmod{n}$, then we start working backward from there. First, we look at the number $a^{2^{k-1}m}$ preceding it. That number is a square root of 1, and if it's anything other than 1 or -1, we know immediately that n can't be prime. In general, we look back to *the last number in the sequence other than 1*: since the number coming right after it—its square—is 1, that number will be a square root of 1, and if it's anything other than -1, we can conclude that n is not a prime.

Like the Fermat test, the Miller-Rabin test seemingly can only give negative results, not positive. But there's a key difference. Each time we run the Miller-Rabin test on a composite number n—that is, pick a number a and raise it to the $(n-1)^{\text{st}}$ power mod n—we might get a result other than 1; in that case, of course, we conclude immediately that n is not prime. But, if n is composite and odd (we're assuming n is odd here; we don't need special instruction for testing even numbers to see if they're prime) we've seen that there are at least four square roots of 1 in arithmetic mod n, meaning at least three square roots of 1 other than 1 itself. Which means that even if $a^{n-1} \equiv 1 \pmod{n}$, the odds that the square root of 1 we find by looking back up the list is other than -1 is at least two to one. The likelihood, therefore, of a false positive on the Miller-Rabin test, therefore, is less than one in three. So if we do, say, 100 iterations of the Miller-Rabin test on a number n, the odds that n will pass the test each time but not be prime is less than 1 in 3^{100}, or somewhat less likely than winning the lottery jackpot every day for a month. That's good enough for us.

Exercise 23.9.1 Try the Miller-Rabin test on the number $n = 561$ using two other witnesses: $a = 4$ and $a = 5$.

Exercise 23.9.2 Try the Miller-Rabin test on the number $n = 1105$ using the two witnesses $a = 3$ and $a = 4$.

Chapter 24

Generators, Roots, and Passwords

We've studied many number systems, and discussed even more. But the one that is without question the most fundamental—that has always exerted the deepest fascination for mathematicians and nonmathematicians alike—is the one we learn first: the counting numbers.

We might ask, in turn: What is the most fundamental property of the counting numbers? Probably most people—at least among those who could be induced to express an opinion in the first place—would say it's *unique factorization*: the fact that every number is a product of primes, and that that expression is unique. It's as central to our understanding of number as the identification of elements is to chemistry.

But unique factorization is in turn a reflection of a basic fact: that while the natural numbers need only one generator with respect to addition, *they require infinitely many generators with respect to multiplication.* In other words, if we have only the operation of addition, we can start with the number 1 and generate all other numbers by adding it to itself. But if we can use only multiplication, there is no analogous generator: not all nonzero numbers are powers of a single number, and in fact the first thing we proved about the natural numbers is that it requires infinitely many numbers—the primes—to generate them multiplicatively.

Now, we've spent the second half of this book talking about other number systems. And our attitude, as we suggested at the outset, has always been one of general curiosity. What aspects of the number systems we're familiar with, we ask, are universal; and which might be different in other number systems?

Take, for example, the question of generators in arithmetic mod p. We know that the number system arithmetic mod p can be generated by a single number with respect to addition: we get all numbers mod p, just as we do all counting numbers, by adding 1 to itself. (One respect in which arithmetic mod p does differ from the counting numbers is that, while 1 is the only number that generates the counting numbers, arithmetic mod p has many possible generators: if you take any number a relatively prime to p and keep adding it to itself, you get all numbers mod p.) All of which naturally leads us to a very fundamental question:

> Can the number system arithmetic mod p be generated by a single number with respect to multiplication?

In other words,

> Is there a number a such that all nonzero numbers in arithmetic mod p are powers of a?

In this final chapter, we'll answer that question, and talk a little bit about generators for arithmetic mod p. We'll also revisit the question of square roots of 1, and also ask when there are square roots of -1 in arithmetic mod p. Finally, we'll also see a beautiful practical application of the notion of generator in creating online passwords.

One warning, though, before we launch into it: this chapter is a little more fast-paced than the ones before it. Like Chapter 7, it's a sort of "capstone" chapter, both tying together some of the ideas we've been looking at and showing you a little of what might lie ahead. So think of yourself as a runner completing a marathon who finally gets a glimpse of the finish line up ahead; give it one more burst of energy and we'll be done.

Exercise 24.0.1 At the other end of the spectrum: show that the number system of ordinary fractions requires infinitely many generators with respect to addition. What would be a good system of generators?

24.1 Generators

Suppose we're given a number p. We'll say that a number a is a *generator* for arithmetic mod p if every number in arithmetic mod p (other than 0, of course) is a power of a.

Notice that the list of powers

$$a, \quad a^2, \quad a^3, \ldots\ldots, a^{p-2}, \quad a^{p-1}$$

has exactly $p-1$ numbers on it. That means that, if there are no repeats, every nonzero number in arithmetic mod p will appear on the list; and conversely, if there are repeats, we won't get them all. Now, as we've observed, if there's a repeat on the list—that is,

$$a^b \equiv a^c \pmod{p}$$

for some pair of numbers b and c between 1 and $p-1$, with b less than c—then we'll have

$$a^{c-b} \equiv 1 \pmod{p};$$

in other words, some power of a below a^{p-1} will be equal to 1 in arithmetic mod p. Thus, *to say that a number a is a generator for arithmetic mod p is exactly to say that no power of a below a^{p-1} is equal to 1 in arithmetic mod p.*

Let's take up now the question raised above: is there always a generator for arithmetic mod p? As is our usual style, we'll start by gathering data—or rather, recalling it; what we need to examine here are the power tables for arithmetic mod p which we wrote out in Part III.

Actually, let's start out with one we didn't do back in Part III: the table of powers in arithmetic mod 5.

	x	x^2	x^3	x^4
1	1	1	1	1
2	2	4	3	1
3	3	4	2	1
4	4	1	4	1

Looking at the table, the answer is clear: neither 1 nor 4 (which is equal to -1 in arithmetic mod 5) can be a generator, but both 2 and 3 are: in the "2" row, for example, we see all nonzero numbers mod 5, and the same is true of the "3" row.

Note, by the way, that if a number is a generator of arithmetic mod p, so is its reciprocal, since the powers of the reciprocal will just be the powers of the original number in reverse order. In this case, 3 is the reciprocal of 2 in arithmetic mod 5.

Let's move on to powers mod 7. Here's the table:

	x	x^2	x^3	x^4	x^5	x^6
1	1	1	1	1	1	1
2	2	4	1	2	4	1
3	3	2	6	4	5	1
4	4	2	1	4	2	1
5	5	4	6	2	3	1
6	6	1	6	1	6	1

Once more, we have two solutions: taking powers of 3 gives us all nonzero numbers in arithmetic mod 7; and the same is true of 5, which is the reciprocal of 3 (mod 7). Note that none of the other numbers in arithmetic mod 7 works here.

Let's do one more. Here's the table of powers in arithmetic mod 11:

	x	x^2	x^3	x^4	x^5	x^6	x^7	x^8	x^9	x^{10}
1	1	1	1	1	1	1	1	1	1	1
2	2	4	8	5	10	9	7	3	6	1
3	3	9	5	4	1	3	9	5	4	1
4	4	5	9	3	1	4	5	9	3	1
5	5	3	4	9	1	6	3	7	9	1
6	6	3	7	9	10	5	8	4	2	1
7	7	5	2	3	10	4	6	9	8	1
8	8	9	6	4	10	3	2	5	7	1
9	9	4	3	5	1	9	4	3	5	1
10	10	1	10	1	10	1	10	1	10	1

Now this is getting interesting. Looking at the table, we see there are *four* possible generators for arithmetic mod 11: the numbers 2, 6, 7 and 8 all work. Of course, they have to come in reciprocal pairs—here $6 = 1/2$ (mod 11) and $8 = 1/7$ (mod 11)—but unlike the last two cases there's more than one such pair.

Exercise 24.1.1 Look at the table of powers in arithmetic mod 13 you made in Part III. (You did do Exercise 18.4.1, didn't you?) Is there a generator? How many?

Exercise 24.1.2 One more example: find the generators in arithmetic mod 17.

Hint: You can make this a lot easier if you establish first that a number a will fail to be a generator in arithmetic mod 17 exactly when $a^8 \equiv 1 \pmod{17}$.

At this point, more examples would serve only to delay the punchline. The fact is that the evidence you've accumulated by now doesn't lie: it is in fact the case that

> If p is any prime number, there is a generator for arithmetic mod p.

As usual, this raises additional questions: how do we know for sure that there's always a generator for arithmetic mod p? Can we say how many? And what use can we make of this? For the first, we're going to have to beg off: we really don't have the tools to give a proof of this fact. (And it does require some: in fact, Euler, in his proof of this fundamental result, created many of the basic tools of modern algebra.) As for the second and the third, we'll get to those in the following two sections.

24.2 More on Square Roots

When we first talked about number systems back in Section 14.3, we talked about the real (decimal) numbers, and the complex numbers. The difference was simple: in the real number system, there is no square root of -1, while in the complex number system there is; indeed, the complex number system is exactly the number system formed by starting with real numbers and demanding that there be such a square root. Another distinction is that in the real number system, in some sense half the nonzero numbers (the positive ones) have square roots and half don't; in the complex number system they all do.

Well, we might ask: which of these properties are shared by our new number systems, arithmetic mod p?

One of these two questions is relatively easy to answer: we'll show you now that in arithmetic mod p, *exactly half the nonzero numbers have square roots*. The key is the fact we saw in Section 23.8 that there are only two square roots of 1 in arithmetic mod p. Now suppose that x is any nonzero number in arithmetic mod p, and suppose that a, b and c are any three square roots of x: that is, suppose

$$x = a^2 = b^2 = c^2$$

for some numbers a, b and c in arithmetic mod p. We ask: what can the ratios

$$\frac{a}{a} \quad \frac{b}{a} \quad \text{and} \quad \frac{c}{a}$$

be? Of course $a/a = 1$, but what about the others?

Well, think about it: if $a^2 = b^2 = c^2$, then when we square the ratios we get

$$\left(\frac{b}{a}\right)^2 = \frac{x}{x} = 1$$

and likewise

$$\left(\frac{c}{a}\right)^2 = \frac{x}{x} = 1.$$

In other words, the ratios b/a and c/a are all square roots of 1 in arithmetic mod p. So either b/a or c/a is equal to 1, in which case $b = a$ or $c = a$; or else $b/a = c/a = -1$, in which case $b = c$.

The conclusion, in other words, is that no nonzero number can have more than two square roots in arithmetic mod p. What's more, if a is a square root of x in arithmetic mod p, then so is $-a$; so the bottom line is that in arithmetic mod p, *every nonzero number has either exactly two square roots or none at all.*

Now, look at the list of squares of nonzero numbers in arithmetic mod p:

$$1^2, 2^2, 3^2, 4^2, \ldots\ldots, (p-3)^2, (p-2)^2, (p-1)^2.$$

Every number that appears on this list has a square root, obviously; and those that don't, don't. But there are $p - 1$ terms on this list, and we've just seen that every number that appears on the list appears exactly twice! Thus the number of numbers with square roots in arithmetic mod p is $\frac{p-1}{2}$, or exactly half the total number of nonzero numbers in arithmetic mod p.

Let's turn now to the second of the two questions we raised at the beginning of this section: *is there a square root of* -1 *in arithmetic mod* p*?*

This is a beautiful example of the sort of question that can be completely baffling when you first ask it, but that becomes remarkably clear if you just look at it the right way. We could ask directly whether the equation

$$x^2 \equiv -1 \pmod{p},$$

has any solutions; or, equivalently, whether any whole number in the sequence

$$p-1, \quad 2p-1, \quad 3p-1, \quad 4p-1, \ldots$$

is a square, but if you can see the answer to this right off the bat you're doing better than we are. On the face of it, this looks like a hard question.

But it's not! In fact, if we think of it in terms of generators, the answer just pops out. We'll show you how in the next section.

24.3 Square Roots of -1

If someone asked you to list all the nonzero numbers in arithmetic mod 13, you wouldn't hesitate. "1, 2, 3, 4," you'd say, and so on up to "10, 11 and 12." Which is natural enough: this is the order in which the numbers appear when we take the number 1 and add it to itself over and over.

But we've seen that the nonzero numbers in arithmetic mod 13 can also be gotten by taking a single number—a generator—and multiplying it by itself over and over. For example, 2 is a generator for arithmetic mod 13, so we can list the nonzero numbers as

$$
\begin{aligned}
2^1 &\equiv 2 \pmod{13}\\
2^2 &\equiv 4 \pmod{13}\\
2^3 &\equiv 8 \pmod{13}\\
2^4 &\equiv 3 \pmod{13}\\
2^5 &\equiv 6 \pmod{13}\\
2^6 &\equiv 12 \pmod{13}\\
2^7 &\equiv 11 \pmod{13}\\
2^8 &\equiv 9 \pmod{13}\\
2^9 &\equiv 5 \pmod{13}\\
2^{10} &\equiv 10 \pmod{13}\\
2^{11} &\equiv 7 \pmod{13}\\
2^{12} &\equiv 1 \pmod{13}
\end{aligned}
$$

This ordering of the numbers in arithmetic mod 13 is exactly analogous to the standard 1, 2, 3, . . . , 11, 12; we're just using the operation of multiplication to generate the numbers rather than addition. This is, in a sense, as natural a way to order the numbers as the first; the fact that most people wouldn't consider it so is basically a reflection of the fact that, to most of us, addition is a more elementary operation than multiplication.

Now, what would be the point of writing out all the numbers of arithmetic mod 13 in this way? Well, just as the standard order reflects how the numbers behave with respect to addition, this ordering makes transparent how they behave with respect to multiplication. For example, if we put the numbers in the standard order

$$1, 2, 3, 4, 5, 6, 7, 8, 9, 10, 11, 12$$

then the sum of the third and fifth elements on the list is simply the eighth; if we put them in the multiplicative order

$$2, 4, 8, 3, 6, 12, 11, 9, 5, 10, 7, 1$$

then the *product* of the third and fifth elements on the list is simply the eighth.

In particular, *writing the numbers of arithmetic mod 13 in this way makes apparent which numbers have square roots*. For example, if we ask whether 10 has a square root in arithmetic mod 13, the answer may not be immediately apparent; but if we know that 10 is 2^{10} in arithmetic mod 13, it's clear: we have

$$2^{10} = (2^5)^2$$

so 2^5, or 6, is the square root of 10. In the same way, we see that all the even powers of the generator 2 have square roots; and only the even powers; so that *the numbers with square roots in arithmetic mod 13 are exactly the even powers of the generator 2, which comprise every other number on this list.*

The same idea applies, moreover, if we replace the modulus 13 by any prime number p, and the number 2 by any generator a for arithmetic mod p: every nonzero number in arithmetic mod p is a power of a, and the ones that have square roots are exactly the even powers.

Having said this, let's return to the question of whether or not there's a square root of -1 in arithmetic mod p. We said the answer would be plain; it's time to back that up.

To start with, let's suppose that a is a generator for arithmetic mod p. Then we can write all the nonzero numbers in arithmetic mod p as the sequence

$$a, \quad a^2, \quad a^3, \ldots\ldots, a^{p-3}, \quad a^{p-2}, \quad a^{p-1} \equiv 1$$

and we know that the numbers that have square roots in arithmetic mod p are just the even terms in this sequence. So the question becomes: where does the number -1 appear in this sequence?

And we know the answer to that, too: -1 comes exactly in the middle of the sequence. After all, -1 is a square root of 1, and 1 is just the $(p-1)^{\text{st}}$ power of a, so -1 must be a raised to half that power. From the other direction, if we take the number

$$b \equiv a^{\frac{p-1}{2}}$$

and square it, we see that

$$b^2 \equiv \left(a^{\frac{p-1}{2}}\right)^2 \equiv a^{p-1} \equiv 1.$$

So b is a square root of 1, and it's not 1 itself, so it must be -1. In short, if a is any generator of arithmetic mod p, it must be the case that

$$-1 \equiv a^{\frac{p-1}{2}} \pmod{p}.$$

This beautiful formula is due to Euler.

Now, does -1 have a square root? The answer is staring us in the face: it does exactly when the exponent $\frac{p-1}{2}$ is even! OK, then, when is $\frac{p-1}{2}$ even? The answer is, when $p-1$ is divisible by 4; that is, when $p \equiv 1 \pmod{4}$. All in all, the answer to our question is expressed in the statement

If p is any odd prime number, there is a square root of -1 in arithmetic mod p exactly when $p \equiv 1 \pmod{4}$.

In other words, there are square roots of -1 in arithmetic mod 5, 13, 17, and 29; but not in arithmetic mod 7, 11, 19 or 23.

Exercise 24.3.1

1. Exactly one of the numbers 2, 3 and 5 is a generator for arithmetic mod 31. Which one is?
2. Express 30 as a power of the generator you picked in arithmetic mod 31.
3. Do there exist square roots of -1 in arithmetic mod 31?

Exercise 24.3.2 Find the square roots of -1 in arithmetic mod 5, 13, 17, and 29.

Exercise 24.3.3

1. How many square roots of 1 are there in arithmetic mod 15?
2. How many square roots of 1 are there in arithmetic mod 67?
3. How many square roots of -1 are there in arithmetic mod 675?
4. How many square roots of -1 are there in arithmetic mod 61?

Exercise 24.3.4 For which prime numbers p do there exist cube roots of 1 other than 1 itself?

24.4 Congruence Classes of Primes

"Whoa!," you might be thinking after that last section, "Congruence classes of prime numbers! That's something we never talked about." You're right, we never did: we talked about congruences in Chapter 16, and about the distribution of primes in Chapters 10 and 23, but we never talked about the two notions in conjunction with each other. Well, now it's come up and we might as well take a moment out and talk about some of the issues that arise.

To start with, let's discuss primes mod 4. Except for 2, every prime is odd. So we can say there are two kinds of primes: those congruent to 1 (mod 4) and those congruent to 3 (mod 4); and we've seen that at least in one respect the two kinds behave differently. The question naturally comes up, then: how many of each kind are there?

Well, let's take a look. We can list the odd primes in order

$$3, \underline{5}, 7, 11, \underline{13}, \underline{17}, 19, 23, \underline{29}, 31, \underline{37}, \underline{41}, 43, 47, \underline{53}, 59, \underline{61}, 67, 71, \underline{73}\ldots,$$

with the primes congruent to 1 (mod 4) underlined. Now imagine we're taking a vote: each prime in turn will say whether it's congruent to 1 (mod 4) or congruent to 3 (mod 4), and we'll keep a tally of the results. For example, after the first 10 votes it's six to four in favor of the congruent to 3 (mod 4) party; after the first 20 votes it's 11 to 9, and so on.

Now, one issue here is that this election will go on forever: there's no Supreme Court to stop it. But it could still have a meaningful outcome. For example, it could be a landslide: one side might get only finitely many votes, which would mean that after a certain point in the balloting *all* the votes would be for the other side. Or we could look at the percentages after 100 votes, after 1,000 votes, after 10,000 votes and so on; if, as the number of votes counted increased, the percentages got closer and closer to some fixed percentage, we could call that the result.

So: what's the outcome? Well, as far as percentages go, it's a dead heat. The 19th century mathematician Dirichlet, in fact, proved that as you count more and more votes, the percentage of votes for each side gets closer and closer to exactly 50%. Not uniformly, necessarily—there may be runs for one side or the other that will momentarily sway the totals—but from a certain point on, the percentages will stay between 49.9% and 50.1%; from another point farther on, they'll stay between 49.99% and 50.01%, and so on.

In fact, Dirichlet proved much more. If you pick any modulus n, you can similarly ask how the primes are distributed among the various congruence classes mod n.

One thing you can certainly say is that except for the (finitely many) primes that divide n itself, every prime will be relatively prime to n; and so its congruence class mod n will be one of the $\phi(n)$ congruence classes relatively prime to n. But how are they distributed among these classes? Dirichlet showed that in fact they are evenly distributed: that is, for any n and any number k relatively prime to n, the fraction of all primes that are congruent to $k \pmod{n}$ is $1/\phi(n)$, in the sense of the last paragraph. In particular, Dirichlet shows that the election is certainly not a landslide: as a consequence of his work, we see that

> For any number n and any number k relatively prime to n, there are infinitely many primes congruent to $k \pmod{n}$.

Exercise 24.4.1 Excluding 3, there are 24 primes below 100, all of which are either congruent to 1 (mod 3), or congruent to 2 (mod 3). How many of each type would you expect there to be? Now count the actual numbers and compare.

Exercise 24.4.2 Excluding 5, there are 45 primes below 200, all of which are either congruent to 1, 2, 3 or 4 (mod 5). How many of each type would you expect there to be? Again, count the actual numbers and compare.

24.5 Counting Generators

We said at the beginning of this chapter that we'd show you how many generators there are for arithmetic mod p. It's really not so hard, if we just start by assuming the existence of one generator a. All we do is line up the nonzero numbers mod p in the multiplicative order

$$a, \quad a^2, \quad a^3, \ldots\ldots, a^{p-3}, \quad a^{p-2}, \quad a^{p-1}$$

and ask: which powers of a are also generators?

Well, suppose we start with the number a^k. The powers of a^k are just the powers of a with exponent divisible by k:

$$a^k, \quad a^{2k}, \quad a^{3k}, \quad a^{4k}, \ldots$$

and so on. The question is, when does every power of a appear on this list? For that matter, we could ask, when does a itself appear on this list? After all, if a^k is a generator, then some power of it must be equal to a; and conversely, if some power $(a^k)^l$ of a^k is equal to a, then we can get every power of a by raising that in turn to powers: for example, we get a^2 as

$$a^2 = \left((a^k)^l\right)^2 = (a^k)^{2l}.$$

The point is, a^k *will be a generator exactly when a itself is a power of* a^k.

OK, then, when does this happen? Well, the powers of a that are equal to a itself in arithmetic mod p are just the powers whose exponent is 1 more than a multiple of $p-1$: by Fermat

$$a^{l(p-1)+1} \equiv (a^{p-1})^l \cdot a \equiv a \pmod{p}.$$

So in order for a^k to be a generator, some multiple of k has to be equal to 1 more than a multiple of $p - 1$, which is to say we have to be able solve the equation

$$m \cdot k = l \cdot (p - 1) + 1,$$

in whole numbers m and l. Now, solving this equation is tantamount to expressing 1 as a combination of k and $p - 1$, and this is something we can do exactly when k is relatively prime to $p - 1$. In sum, then: *if a is a generator of arithmetic mod p, then a^k is also a generator exactly when k is relatively prime to $p - 1$.*

In particular, this tells us how many generators there are: it's just the number of numbers between 1 and $p - 1$ that are relatively prime to $p - 1$; that is, $\phi(p - 1)$.

Exercise 24.5.1 How many generators are there for arithmetic mod 23? Can you find them?

24.6 "Swordfish"

When last we left Alice and Bob (and Eve) in Section 22.3, Alice and Bob were enjoying a nice extended conversation, with Eve listening in but unable to decode the messages. Each day, Alice contacts Bob, who constructs the key for the day: that is, he multiplies two large primes p and q to get a modulus n; he chooses a number k relatively prime to $\phi(n)$, and instructs Alice to encode her messages to him by converting them to numbers and raising those numbers to the kth power in arithmetic mod n. (Alice likewise constructs a code for Bob to use in sending messages to her.)

But one day Eve has an idea. She can't break the code that Bob devises. But *what if she intercepts Alice's initial message to Bob, and pretends to be Bob herself?* Bob will never even know that Alice is trying to contact him; instead, Eve will make up the codes and so will be able to understand everything Alice says.

How can Alice and Bob thwart Eve's evil plan? In other words, each day when Alice fires up her computer and initially contacts Bob to set up the conversation, *how does she know it's really Bob she's talking to?*

What they need, clearly, is a password: something only they know, so that when Alice first contacts Bob she can ask him to give the password, and if he does she can be confident it's really him she's talking to. But how can they exchange a password when *Eve is listening to everything they say?*

It seems our heroes are really in a pickle! But once more, modular arithmetic saves the day. We'll see in this section how we can use the special properties of arithmetic mod p to create what are called *signatures*. Using these, two people can exchange a password that will allow them to identify each other in the future, in such a way that a third party who overhears every word they exchange won't know the password.

How does this work? Well, like the public-key cryptosystems we described earlier, these are based on the fact that there are operations we can carry out easily in modular arithmetic but that we can't undo, or reverse, in reasonably short time (e.g., less than the lifetime of the universe). Specifically, suppose first of all we pick a large prime number p—say three hundred digits long. Next, suppose that a is a generator for arithmetic mod p. Then we know that every number b in arithmetic mod p is a power of a, *but we can't actually say what power it is*. We know if that if we could make a list of the powers of a

$$a, \quad a^2, \quad a^3, \ldots\ldots, a^{p-3}, \quad a^{p-2}, \quad a^{p-1}$$

then b would be on it someplace, but there's no straightforward way of telling in advance where; and even just reading through such a list to locate k would take longer than the lifetime of the universe. In short: we can raise a to any power we like instantly; but if we're given a number b we can't readily tell what power of a it is.

Given this, here's how Alice and Bob come up with a password:

First step (Public). To start with, either Alice or Bob—it doesn't matter who—chooses a large prime p and a generator a for arithmetic mod p. These numbers are exchanged publicly; Alice, Bob and Eve all know them.

Second step (Private). Next, Alice secretly picks a number k between 1 and $p-2$, and likewise Bob picks a number l between 1 and $p-2$. These numbers are secret.

Third step (Public). Now, Alice calculates the power a^k (mod p), and likewise Bob calculates a^l (mod p), and they share these numbers publicly: again, Alice, Bob and Eve all know the numbers a^k and a^l (mod p).

Fourth step The payoff: the password is a^{kl}. Alice can figure out the password: she knows k (she picked it) and she knows a^l (Bob told her), so she can raise a^l to the kth power to obtain

$$(a^l)^k \equiv a^{kl}.$$

Likewise, Bob knows l and he knows a^k, so he can calculate

$$(a^k)^l \equiv a^{kl}.$$

But Eve, who knows p, a, a^k and a^l but doesn't know either k or l, has no way to figure out what a^{kl} is. So the next time Alice and Bob want to talk, and Alice wants to be sure it's Bob she's talking to, she can ask him for the password—he's the only one in the world besides her who knows it.

Pretty slick, huh? Alice and Bob jointly create a bit of knowledge that is known only to the two of them, and even someone who has overheard every communication between them isn't privy to the secret. It beats "swordfish," anyway. And, while it may seem esoteric, in fact these signatures (or variants of them) are used every day to make sure electronic communication is secure.

Two remarks here. First, each time Alice and Bob talk—once they've made sure it really is the other they're talking to—they need to generate a new password to use the next time they want to contact each other. Also, note that the password Alice and Bob have created can be used to reassure Alice that it's really Bob she's talking to; if Bob needs to be similarly reassured that it's Alice on the other end of the line, they'll have to generate two passwords each time.

It's time now to leave Alice and Bob (and Eve) to their communication, secure or otherwise. It's:

24.7 Time to Say Goodbye

We started this book with a crack about the opening chapters of most math books; we might as well close it similarly.

Most math books, in our experience, suffer from a tendency to pile on at the end. It's a natural, understandable tendency: there's so much more to do! So many fascinating questions just over that ridge, that we're so close to being able to answer! It's so tempting to put them in; after all, we say to ourselves, the reader can just skip them if he's not interested. And, of course, the more topics the book covers the easier it'll be to sell.

But there's a price to be paid for tacking on all that stuff at the end. Instead of arriving at the end of the book with a sense of accomplishment and completion, the reader has his nose rubbed repeatedly in the fact that there's so much more out there—that what's been covered is in the end only a tiny fraction of all there is to know. In the end the reader puts down the book not so much satisfied as exhausted.

Now, you might say we've already committed this sin, that this book should have ended a chapter or two ago. You might be right, but in any event we're not going to compound the error. If you've stayed with us till now, you know a tremendous amount of real mathematics—more than all but a small fraction of people walking around. More importantly, perhaps, you have a sense of what mathematics is: a separate universe that nonetheless pervades and underlies the world we live in. You've done a good job; and now it's time for us to say goodbye and for you to close the book and get on with your life.

HEY, I READ THIS BOOK!
I THOUGHT THE AUTHOR GOT OFF ON SOME ODD TANGENTS FOR A LOT OF IT.
BUT OVERALL, MOST OF THE POINTS WERE GOOD, AND IT RAISED SOME INTERESTING PARALLELS. I LIKED HOW IT CAME FULL CIRCLE IN THE LAST CHAPTER.
JASON, IT'S A MATH BOOK!
2πr! WOO! WHAT AN ENDING!
© 2001 Bill Amend / Distributed by Universal Press Syndicate
5-18
AMEND

INDEX

N

P

R

S

T

V

W